Mastering Simulink

Mastering Simulink

James B. Dabney

and

Thomas L. Harman

University of Houston–Clear Lake

PEARSON

Prentice
Hall

Upper Saddle River, New Jersey 07458

Library of Congress Cataloging-in-Publication Data

Dabney, James.
   Mastering Simulink / James B. Dabney and Thomas L. Harman.
     p. cm.
   ISBN 0-13-142477-7
   1. Computer simulation. 2. SIMULINK. I. Title: Mastering Simulink. II. Harman,
Thomas L., 1942- III. Title.

   QA76.9.C65D34   2004
   003$'.85'0113536$-dc22

2003059516

Vice President and Editorial Director, ECS: *Marcia J. Horton*
Publisher: *Tom Robbins*
Associate Editor: *Alice Dworkin*
Vice President and Director of Production and Manufacturing, ESM: *David W. Riccardi*
Executive Managing Editor: *Vince O'Brien*
Managing Editor: *David A. George*
Production Editor: *Colleen Franciscus*
Director of Creative Services: *Paul Belfanti*
Art Director: *Jayne Conte*
Cover Designer: *Bruce Kenselaar*
Cover Art: *Robert Tinney*
Art Editor: *Greg Dulles*
Manufacturing Manager: *Trudy Pisciotti*
Manufacturing Buyer: *Lynda Castillo*
Marketing Manager: *Holly Stark*

© 2004 by Pearson Education, Inc.
Pearson Prentice Hall
Pearson Education, Inc.
Upper Saddle River, NJ 07458

Pearson Prentice Hall® is a trademark of Pearson Education, Inc.

MATLAB, Simulink, Stateflow, Handle Graphics, Real-Time Workshop, and xPC Target are registered
trademarks of The MathWorks, Inc.

The author and publisher of this book have used their best efforts in preparing this book. These efforts
include the development, research, and testing of the theories and programs to determine their
effectiveness. The author and publisher make no warranty of any kind, expressed or implied with regard
to these programs or the documentation contained in this book. The author and publisher shall not be
liable in any event for incidental or consequential damages in connection with or arising out of the
furnishing, performance or use of these programs.

Printed in the United States of America
10 9 8 7 6 5 4 3 2 1

ISBN 0-13-142477-7

Pearson Education Ltd., *London*
Pearson Education Australia Pty. Ltd., *Sydney*
Pearson Education Singapore, Pte. Ltd.
Pearson Education North Asia Ltd., *Hong Kong*
Pearson Education Canada, Inc., *Toronto*
Pearson Educación de Mexico, S.A. de C.V.
Pearson Education—Japan, *Tokyo*
Pearson Education Malaysia, Pte. Ltd.
Pearson Education, Inc., *Upper Saddle River, New Jersey*

To Beth

Brief Contents

Contents

Preface

We intend for this book to serve as a tutorial for new users of Simulink and as a reference for experienced users. The book covers all of the important capabilities of Simulink, including subsystems, masking, callbacks, S-Functions, and debugging. The book is meant to be used with Simulink 5 and subsequent revisions. The examples were produced with Simulink Version 5.0. Simulink is a programming language specifically designed for simulating dynamical systems. Therefore, in order for you to use Simulink effectively, you should have the appropriate mathematical preparation. We assume you have a good understanding of the concepts usually covered in the introductory courses in calculus and differential equations. However, as many new users of Simulink may be unfamiliar with block diagram notation, we included a chapter that introduces the notation.

USING THE BOOK

Here, we offer suggested reading sequences for new users of Simulink, for users experienced with a previous version of Simulink, and for advanced users ready to take advantage of all of the power of Simulink.

New Users

It is possible to model fairly complex systems with basic proficiency with Simulink. The fastest way to gain this basic proficiency is to adhere to the following sequence:

1. If you are new to block diagrams, read Sections 2.1 and 2.2. These sections introduce block diagram notation and illustrate using block diagrams to model scalar continuous systems.

2. Carefully work through all of the examples in Chapters 3 and 4 to master the mechanics of building and running models.

3. Read Sections 5.1, 5.2, and 5.4 and experiment with the examples. After completing this material, you should be comfortable building and running models of scalar continuous systems.

4. As you gain proficiency with Simulink, complete Chapter 2, then work through the rest of Chapter 5 and Chapter 6.

5. If you have access to Stateflow, work through Chapter 14.

6. If you have access to Real-Time Workshop, work through Sections 15.1 through 15.3 and read Section 15.4. If you have access to xPC, also work through Section 15.4.

Experienced Users

If you are experienced with a previous version of Simulink, or, if you are a new user, after you have acquired basic proficiency, we suggest you proceed as follows:

1. Read Section 3.4. The new help system provides detailed online documentation for all Simulink blocks. We believe that you will find the help system to be easy to use and to be a real time saver.

2. Review Chapter 4. The Simulink user interface has many improvements over the previous version of Simulink. Pay particular attention to Section 4.8 concerning selecting and configuring a solver.

3. Scan Chapters 5 and 6. Pay particular attention to Section 5.2.1.

4. Read Sections 7.1 and 7.2, then work through Sections 7.3 and 7.4 in detail. Learning to use conditionally executed subsystems will allow you to build efficient models.

5. Read Chapter 8, even if you don't plan to use the analysis capabilities right away. You may well discover that the analysis tools will make your use of Simulink much more productive.

6. Read Chapter 12 carefully. The new debugging features can save lots of time.

7. Review Chapter 13. An understanding of the numerical issues can allow you to build models that are faster and more accurate.

Advanced Users

If you are already experienced with Simulink 5, we suggest you proceed as follows:

1. Scan Chapter 4 to review the basics of model building, and scan Chapter 13 to review the numerical issues.

2. Read Chapters 7 and 8 to review subsystems, masking, and Simulink analysis tools.

3. If you intend to build graphical user interfaces or interactive animations, read Chapter 9 and study the examples.

4. Review Chapter 10, particularly Sections 10.1 through 10.4. Even if you don't need to use S-Functions right away, understanding the capability will allow you to recognize situations in which S-Functions are appropriate.

5. Review Chapter 11 and experiment a little with the Animation Toolbox and Dials and Gauges, if available.

ACKNOWLEDGMENTS

We wrote this book with a tremendous amount of help. The MathWorks was very supportive. In particular, we wish to acknowledge Jim Tung, Rick Spada, Loren Dean, Kevin Kohrt, Liz Callanan, Mehran Mestchian, Ben Hinkle, and Naomi Fernandes. We wish to thank Prentice Hall for ably managing production.

Finally, we wish to acknowledge the continuous support and encouragement provided by our publisher, Tom Robbins, over the last seven years. Tom provided a great deal of helpful advice, solved every administrative problem we encountered, and made sure we had all the resources we needed to complete the project.

1

Introduction

In this chapter, we will explain what Simulink is and why it's important. Next, we will look at some industry applications of Simulink. Then, we'll discuss two important approaches to using Simulink: rapid prototyping and rapid application development. We'll then discuss factors you should consider when selecting a simulation environment. We will conclude Chapter 1 with a preview of the remaining chapters.

1.1 WHAT IS Simulink?

Simulink is an extension to MATLAB that allows engineers to rapidly and accurately build computer models of dynamical systems using block diagram notation. Using Simulink, it is easy to model complex nonlinear systems. A Simulink model can include continuous and discrete-time components. Additionally, a Simulink model can produce graphical animations that show the progress of a simulation visually, significantly enhancing understanding of the system's behavior.

In the past, a common approach to developing a computer model of a dynamical system was to start with a block diagram. Next, the block diagram was translated into the source code of a programming language. This practice involved duplication of effort, as the system and controller had to be described twice—once in block diagram form and then again in the programming language. It also introduced the risk that the translation from block diagram to source code may be inaccurate. A difficult problem in debugging a control system design was determining the location of an error. The error could be in the design (block diagram world), in the program (programming language world), or in the translation from block diagram to program. With Simulink, the duplication of effort in rewriting the model in a programming language is eliminated, and the fact that the "program" is the block diagram eliminates the risk that the program may not accurately implement the block diagram.

The potential productivity improvement realized from this block diagram approach to programming is dramatic. As an example, consider the simple spring-mass system of Figure 1.1. A Simulink model of this system is shown in Figure 1.2. Table 1.1 provides a comparison of programs modeling this system in 8086 assembly language, using 16-bit integer arithmetic and simple Euler integration, in FORTRAN using floating-point arithmetic, in MATLAB using matrix arithmetic, and in Simulink. For the Simulink program, we list the number of blocks rather than program statements, and we include the number of mouse clicks with the number

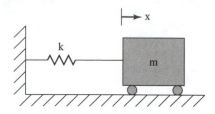

FIGURE 1.1: Spring-mass system

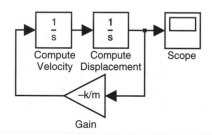

FIGURE 1.2: Simulink model of spring-mass system

TABLE 1.1: Program Size Comparison

Programming Language	Lines of Code (Blocks)	Approximate Keystrokes
8086 Assembly Language	92	1540
FORTRAN	14	240
MATLAB	3	90
Simulink	4	25

of keystrokes. As there is a strong correlation between the number of lines of code required to produce a program and the time required to write the program, it is clear that Simulink provides the opportunity for significant productivity improvements.

1.2 Simulink EXAMPLES

Simulink is now widely used in industry. Two examples of Simulink use in industry are the modeling of a steel rolling plant and the modeling of internal combustion engines. We'll look at examples of each application next.

1.2.1 Steel Rolling Plant Modeling

Our first example of industrial use of Simulink is in the simulation of a steel rolling plant [1]. The steel rolling process begins with flat-cast 25-ton slabs that are fed into the rolling mills at 1200°C (2200°F). The slabs move through the mills at the rate of 20 slabs per hour, exiting the rolls at 10 meters per second. The objective of the plant control system is to maximize the rate of production while maintaining the desired material properties and thickness of the finished product.

It is not economically feasible to experiment with the plant controls to evaluate control strategies to increase production. Consequently, the plant engineers developed a high-fidelity Simulink simulation of the plant and the control system. Using the Simulink model, controller parameters are fine-tuned offline, and new control algorithms are evaluated. Additionally, the Simulink model is used as a training tool for new engineers. The advantages of Simulink use, in this case, are the ease with which the Simulink model may be modified and the fact that the Simulink model is nearly self-documenting.

1.2.2 Automotive Engine Modeling

Simulation is an important tool in the development of automotive engines. A Simulink model of a port fuel-injected internal combustion engine was developed by Simcar.com [3]. This is a complex, hierarchical model that accounts for nonlinear air, fuel, and exhaust gas recirculation dynamics in the intake manifold, as well as internal process delays. A top-level view of the Simulink model is shown in Figure 1.3.

The engine model is used in several ways. It is used as a non-real-time engine model for testing engine control algorithms and engine sensor and actuator models. It is also used as a real-time model for testing prototype controllers. It may also be

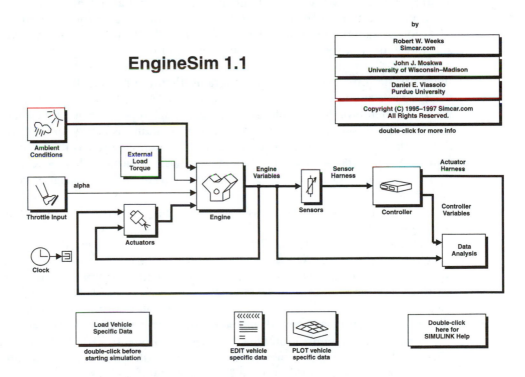

FIGURE 1.3: Simulink automotive engine model

used as an observer embedded within a control algorithm or as an engine subsystem model embedded in a powertrain or vehicle dynamics simulation. It has been found that the ability to rapidly evaluate new strategies in Simulink permits engineers to try ideas in minutes or hours, instead of the days or weeks previously required to test the ideas on an actual vehicle [3].

1.3   Simulink EMPLOYMENT STRATEGIES

So far, we looked at the evolution of engineering programming languages and the improvements in productivity that Simulink offers. In this section we will discuss the two principal strategies for Simulink employment: rapid prototyping and rapid application development (RAD). Rapid prototyping is the application of productivity tools to develop working prototypes of control systems in the minimum amount of time. Rapid prototyping methods are optimized for development speed, rather than execution speed or memory use. RAD is an extension to the rapid prototyping methodology in which the final computer program is the Simulink model or is derived automatically from the Simulink model.

1.3.1   Rapid Prototyping

Digital control systems are built using general-purpose computers or embedded microprocessors. In both cases, it is very expensive and time-consuming to develop and debug controller software using the target physical system. In some cases, such as aircraft autopilots, it would not only be expensive but also dangerous to attempt to develop the controller software using the target computer and physical system. The most common approach to developing controller software is to first develop a software model of the plant to be controlled, and then use this model to develop a simulation of the control system. A common development strategy is to design the controller on paper using block diagrams, then to implement the design in the desired programming language, and integrate the controller with the plant model. Next, parameters of the controllers are tuned through repeated runs of the combined controller-plant simulation. If the controller proves unsatisfactory, or if alternative designs are to be evaluated, the design and programming cycle is repeated. Rapid prototyping provides a means to streamline the design of a controller by delaying the translation from block diagram to computer program until the design is complete and verified.

The rapid prototyping process begins with the development of a simulation of the physical system to be controlled. The simulation can be a Simulink model, or it can be a model developed in MATLAB or C. Next, the controller is developed using Simulink and perhaps MATLAB toolboxes, such as the Optimization Toolbox. This phase is really a combination of the requirements and design phases of traditional software development cycles. During the prototyping phase, a detailed set of test cases is executed to verify that the design satisfies all performance requirements. Alternative designs may be tried, and controller parameters may be tuned.

Once the Simulink prototype has been thoroughly tested, the controller software is coded in a programming language compatible with the target computer. The program is then loaded into the target computer, and tests are run using the operational controller to control the computer model of the plant. This testing verifies that the operational program correctly implements the design that was verified using the Simulink model. Last, the operational controller is tested using the actual plant.

There are several advantages to the rapid prototyping approach. One advantage is that the time-consuming process of developing controller software using a traditional programming language is performed only once. A second advantage is that alternative designs may be tried relatively inexpensively, promoting innovation. Finally, having the design verified before programming begins facilitates debugging during the programming phase, as the likelihood of errors being in the design rather than in the program is low.

1.3.2   Rapid Application Development

RAD extends the rapid prototyping idea by eliminating programming in a traditional language. In cases in which the controller is a computer compatible with Simulink, Simulink can be used as the final programming language. In the steel rolling mill application discussed earlier, the Simulink model is used operationally as the training simulation for new engineers.

Often, the Simulink model of the controller is not suitable for direct deployment. In some applications, the Simulink controller does not respond fast enough. In many applications, the controller computer is an embedded microprocessor that is not compatible with Simulink. Both of these types of applications are candidates for automatic code generation using The MathWorks' Real-Time Workshop. Real-Time Workshop automatically converts the Simulink block diagrams into C language programs that may be compiled for the target computer. The automatically generated controller code may then be linked with off-the-shelf real-time executive[1] and interface code to produce an operational control system for embedded microprocessors. In applications where the Simulink model is not fast enough, translating the Simulink controller into C using Real-Time Workshop will provide substantial speed improvements, particularly when run on high-speed simulation hardware available from dSPACE, Applied Dynamics, and a number of other vendors.

1.4   SELECTION CONSIDERATIONS

Simulink is a very powerful tool for modeling and control system design, and, like every powerful tool, it is better suited to some tasks than to others. Sometimes, it's better to use other special-purpose tools. For example, although it's possible to model electronic circuits using Simulink, there are special-purpose software packages designed for that specific task. Of course, many programming tasks are best handled with general-purpose programming languages. There are three key questions that must be answered in choosing a tool for an engineering design task:

1. Is the tool suitable for the task? Simulink is designed to model dynamical systems using block diagram notation. If simulation is a significant part of the project, Simulink should be considered.

---

[1] A real-time executive is a special type of computer operating system for computers in control systems. The real-time executive ensures that important tasks are performed on schedule.

2. What are the schedule implications? Time is an important factor in every engineering project, both in the university and in industry. If Simulink is suitable for the task at hand, it can save a significant amount of time.

3. How will the choice of development tools affect the total project cost? It has been estimated that it can cost from 10 to 100 times as much to correct a software error found in operational software than to correct the same error in the design stage. [2] A Simulink model can be thought of as an executable software design document. Thus, whether Simulink is used for rapid prototyping or rapid application development, it has the potential to save significant development costs.

1.5 OUTLINE OF TEXT

This text is intended to supplement the principal text for courses in control systems, dynamics, or simulation, and is also meant to be a tutorial and reference for practicing engineers. Chapter 2 provides an overview of block diagrams for continuous and discrete systems. Chapter 3 provides a quick introduction to basics of model building, followed by detailed coverage in Chapter 4. Chapter 5 discusses modeling continuous systems, and Chapter 6 discusses discrete-time systems. Chapter 7 shows how to model more complex systems using hierarchical block diagrams and custom blocks called *masked blocks*. Chapter 8 discusses the use of Simulink analysis tools, including the linearization and trim functions. Callback functions, discussed in Chapter 9, provide a means to add functionality to Simulink models using MATLAB M-files. S-Functions, discussed in Chapter 10, allow you to create new Simulink blocks using either the MATLAB language or C. Chapter 11 shows how to add animations using S-Functions and the Animation Toolbox. Chapter 12 discusses debugging and the Simulink debugger. Chapter 13 discusses numerical issues of concern in Simulink modeling. Chapter 14 presents a introduction to Stateflow. Finally, Chapter 15 presents a brief introduction to Real-Time Workshop.

REFERENCES

[1] "Simulink Helps BHP Steel Achieve World-Class Production," Application Note (The MathWorks, 1994).

[2] *Software Independent Verification and Validation*, AFSC AFLC Pamphlet 800-5, Department of the Air Force (Washington, DC, 1988), 2.

[3] Weeks, Robert W., and Moskwa, John J., "Automotive Engine Modeling for Real-Time Control Using MATLAB/Simulink," Vehicle Computer Applications: Vehicle Systems and Driving Simulation, SP-1080 (Society of Automotive Engineers, 1995), 123–137.

2

Block Diagrams

This chapter is devoted to an introduction to block diagrams. We will start with primitive blocks used to describe simple continuous systems and progress to discrete and hybrid systems. We will also discuss the state-space concept and development of state-space models from the describing differential and difference equations. For each block discussed, we will show the standard block diagram notation and, where they're different, the equivalent Simulink block. Several examples illustrate the use of the block diagram components and the relationship between the block diagrams and differential and difference equations.

2.1 INTRODUCTION

In Chapter 1, we looked briefly at the uses of Simulink and some example Simulink programs. Before we proceed to explore Simulink, we will introduce block diagrams for continuous and discrete systems. We will examine the most common blocks for both types of systems. More specialized blocks will be introduced in later chapters.

Block diagram notation is a graphical means with which to represent dynamical systems. The notation has been used for many years. While block diagrams originally were used to represent linear time-invariant continuous systems, the notation has been extended to represent complex nonlinear systems containing continuous-time and discrete-time components. A continuous system is a system that may be represented using differential equations. A discrete system is represented using difference equations. A hybrid system contains continuous and discrete components. A typical hybrid system would consist of a continuous process controlled by a digital computer. Today, the hybrid system is the most common.

Block diagrams bear a superficial resemblance to computer program flow-charts, but they are conceptually quite different. A flowchart describes a sequence of operations, so only one block in the flowchart is active at a time. A block diagram describes a set of relationships that hold simultaneously. All blocks in a block diagram may be active at once, so a block diagram may be thought of as representing a set of simultaneous equations. Thus, a computer program that models a dynamical system would evaluate each block in the block diagram repeatedly. (Note that Simulink has the capability, using Stateflow, to add procedural blocks to a Simulink model, but this additional capability is not based on the concepts of block diagrams.)

We will start our discussion of block diagrams with continuous systems. We will look at typical components of a continuous block diagram and their Simulink representations. Next, we'll show how the blocks may be composed to model simple continuous systems. In later sections of this chapter, we'll examine block diagrams for discrete and hybrid systems.

2.2 CONTINUOUS SYSTEMS

Most physical systems are modeled as continuous systems, because they can be described using differential equations. The simplest models are linear and time-invariant. Although few physical systems are truly linear or time-invariant, much useful insight can be gained using linear time-invariant models. Additionally, the blocks used to model such idealized systems are needed to model the behavior of nonlinear and time-varying systems.

Four primitive blocks are used to represent continuous linear systems. These are the *gain block*, the *sum block*, the *derivative block*, and the *integrator*. Any system that can be described using linear differential equations may be modeled using these four primitive blocks. In addition to the four primitive blocks, the transfer function block is frequently used in the modeling of physical systems and controllers.

2.2.1 Gain Blocks

The simplest block diagram element is the gain block (Figure 2.1). The output of the gain block is the input multiplied by a constant. Thus, the block represents the algebraic equation

$$y = kx$$

A simple physical example of a gain block is the force balance of a lever. Referring to Figure 2.2, the lever in equilibrium can be represented by the equation

$$F_{out} = aF_{in}$$

where a is the length ratio,

$$a = l_1/l_2$$

Other physical examples that would be represented using gain blocks are the linear model of an electrical resistor and the linear spring. The equation for the linear resistor is

$$v = RI$$

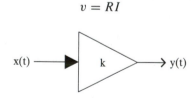

FIGURE 2.1: Gain block

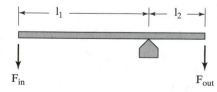

FIGURE 2.2: Lever force multiplication

in which the input is the current I, the gain is the resistance R, and the output is the voltage v. The behavior of a linear elastic spring is represented by the equation

$$F = kx$$

where the input is the spring deflection x, the gain is the spring constant k, and the output is the force F.

2.2.2 Sum Block

The sum block permits us to add two or more inputs. The output of the sum block, shown in Figure 2.3, is the algebraic sum of the inputs. Each input is labeled with a plus sign $(+)$ or a minus sign $(-)$ to indicate whether the input is to be added or subtracted. Thus, the sum block in Figure 2.3 represents the algebraic equation

$$c = a - b$$

A sum block must have at least one input and exactly one output. While there is, in general, no limit to the number of inputs, too many inputs makes the block diagram difficult to read. When the number of inputs to a sum block begins to make the diagram difficult to read, it's better to cascade several sum blocks.

2.2.3 Derivative Block

The derivative block computes the time rate of change of its input. The block represents the differential equation

$$y = \frac{dx}{dt}$$

Shown in Figure 2.4 are two common representations of the derivative block and the Simulink representation. The second representation (the block containing s) is due to the fact that the Laplace transform of the derivative of a function (ignoring

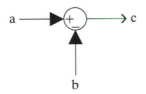

FIGURE 2.3: Sum block

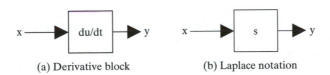

(a) Derivative block   (b) Laplace notation

FIGURE 2.4: Derivative blocks

initial conditions) is

$$\mathcal{L}\left(\frac{dx(t)}{dt}\right) = s\mathcal{L}(x(t))$$

where $\mathcal{L}$ represents the Laplace operator and s the Laplace domain complex frequency variable. We will use the representation of Figure 2.4(a) for consistency with Simulink. A derivative block has a single input and a single output.

A physical example of a system component that could be represented using a derivative block is a dashpot, as depicted in Figure 2.5. The force equation of the dashpot is

$$F = c\frac{dx}{dt}$$

where F is the damping force and c is the damping coefficient.

2.2.4 Integrator

The integrator block computes the time integral of its input from the starting time to the present. An integrator block is shown in Figure 2.6(a), along with an alternate representation in Figure 2.6(b). The integrator block represents the equation

$$y(t) = y(t_0) + \int_{t_0}^{t} x(\tau)d\tau$$

The $1/s$ label is a reference to the Laplace transform representation of integration: recall that integration in the time domain corresponds to multiplication by $1/s$ in the Laplace complex frequency (s) domain. It is now more common to see the $1/s$ representation in block diagrams than the integral sign representation. We will use the $1/s$ notation for consistency with Simulink.

An example of a physical component that could be represented using an integrator block is the capacitor, as shown in Figure 2.7. The input to the block is the current divided by the capacitance, and the output is the voltage across the capacitor plates.

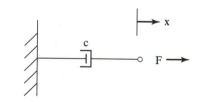

FIGURE 2.5: Dashpot model

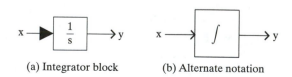

(a) Integrator block   (b) Alternate notation

FIGURE 2.6: Integrator blocks

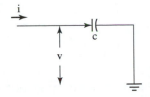

FIGURE 2.7: Capacitor circuit

2.2.5 Simple Physical Models

The primitive block diagram components we've defined so far may be used to model any physical system that may be described completely using linear differential equations. To see how we build block diagrams using these components, consider the simple cart shown in Figure 2.8. Ignoring friction, the equation of motion of this system is

$$\ddot{x} = \frac{F}{m}$$

[Note that we use the dot notation ($\dot{x}$) to represent the time derivatives, and thus, $\ddot{x}$ is equivalent to $\dfrac{d^2x}{dt^2}$.] This system may be represented by the block diagram shown in Figure 2.9. We can expand this block diagram to compute the cart position. In Figure 2.10, we added two integrators. The first computes the cart velocity, and the second computes displacement.

2.2.6 Transfer Function Block

Transfer function notation is frequently used in control system design and system modeling. The transfer function can be defined as the ratio of the Laplace transform of the output to a system (or subsystem) to the Laplace transform of the input, assuming zero initial conditions. Thus, the transfer function provides a convenient

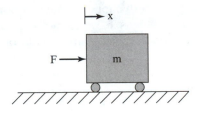

FIGURE 2.8: Cart

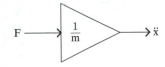

FIGURE 2.9: Block diagram of cart equation of motion

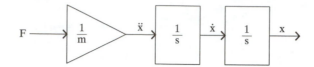

FIGURE 2.10: Block diagram of cart position computation

input–output description of the system dynamics. As we'll see, the transfer function block is a compact notation for a composition of primitive block diagram components.

Consider the spring-mass-dashpot system depicted in Figure 2.11. Ignoring friction, the equation of motion of this system is

$$m\ddot{x} + c\dot{x} + kx = F$$

Taking the Laplace transform and ignoring initial conditions yields

$$ms^2 X(s) + csX(s) + kX(s) = F(s)$$

In Figure 2.12, this system is depicted using primitive block diagram components. The ratio of the Laplace transform of the output $X(s)$ to the Laplace transform of the input $F(s)$ is the transfer function $G(s)$,

$$G(s) = \frac{X(s)}{F(s)} = \frac{1/m}{s^2 + \frac{c}{m}s + \frac{k}{m}}$$

In Figure 2.13(a), the same system in transfer function form is represented. The transfer function representation is more compact, and it is useful in understanding the system dynamics. The denominator of the transfer function is the characteristic equation of the system, and thus, the roots of the denominator are the eigenvalues of the system. (See Harman [2] for a detailed discussion of eigenvalues.)

While the transfer function of a dynamical system is unique, the differential equation that describes a dynamical system is not unique. Thus, if we need access to internal variables, such as $\dot{x}$, we cannot use transfer function notation. For the same reason, if we need to specify initial conditions, we cannot use transfer function notation.

Shown in Figure 2.13(b) is an alternate notation for the transfer function block. This representation treats the transfer function symbolically and is commonly used

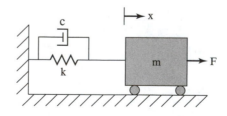

FIGURE 2.11: Spring-mass-dashpot system

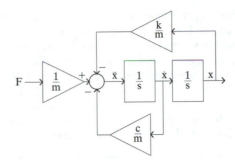

FIGURE 2.12: Block diagram of spring-mass-dashpot system

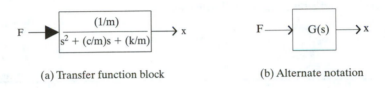

(a) Transfer function block   (b) Alternate notation

FIGURE 2.13: Transfer function representations of a spring-mass-dashpot system

in diagrams with multiple transfer function blocks. A common convention is to represent the transfer function of the plant as $G(s)$ and the transfer function of the controller as $H(s)$.

2.3   STATE-SPACE BLOCK

The *state-space block* is an alternative to the transfer function block. Like the transfer function block, the state-space block provides a compact representation of the system or subsystem dynamics. However, the state-space block also permits us to specify initial conditions and can provide access to internal variables. Another advantage of the state-space block is that it allows us to conveniently model systems with multiple inputs and multiple outputs; whereas the transfer function block accommodates only one input and one output.

2.3.1   The State-Space Concept

Before we explain the definition of the state-space block, we will discuss the concept of state variables, then show how the state-space model of a system is developed.

A *state vector* is a set of *state variables* sufficient to describe the dynamic state of a system. The general form of the state-space model of a dynamical system is

$$\dot{\mathbf{x}} = \mathbf{f}(\mathbf{x}, \mathbf{u}, t) \tag{2.1}$$

where $\mathbf{x}$ is the state vector, $\mathbf{u}$ the input vector, and t time. Equation (2.1) is called the system state equation. We also define the system output to be

$$\mathbf{y} = \mathbf{g}(\mathbf{x}, \mathbf{u}, t) \tag{2.2}$$

Equation (2.2) is called the *output equation*.

Note that we will use lowercase boldface type to distinguish vectors and uppercase boldface type to distinguish matrices.

The so-called *natural state variables* are physical values such as position, velocity, temperature, or electrical current. For a mechanical system, the natural state variables are positions and velocities. For electrical circuits, the natural state variables could be voltages or currents. The natural state variables are not the only set of state variables we can choose. In fact, any independent combination of a valid set of state variables is also a valid set of state variables. In Example 2.1, the selection of state variables for a single-degree-of-freedom system is illustrated.

EXAMPLE 2.1 Pendulum model

Consider the pendulum of Figure 2.14. A natural set of state variables would be the deflection angle and the rate of change of deflection:

$$x_1 = \theta$$

$$x_2 = \dot{\theta}$$

The state-space approach becomes very useful as the complexity of a system increases. Example 2.2 illustrates the selection of state variables for a system with two degrees of freedom.

EXAMPLE 2.2 Two-mass system

The two-mass system in Figure 2.15 has two degrees of freedom, each of which is described by a second-order differential equation, so we need four state variables to specify the dynamic state of the system. A natural set of state variables would be $w_1 = x_1$, $w_2 = \dot{x}_1$, $w_3 = x_2$, and $w_4 = \dot{x}_2$.

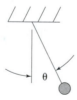

FIGURE 2.14: Pendulum model

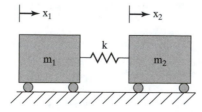

FIGURE 2.15: Two-mass system

2.3.2   Linear Single-Input, Single-Output Systems

The state variable approach is particularly useful in modeling linear systems, because we can take advantage of matrix notation to describe complex systems in a compact form. Additionally, we can compute the system response using matrix arithmetic. In this section, we will use the state-space approach to model linear systems with one input and one output.

The natural state variables of a mechanical system are position and velocity. Consider the spring-mass-dashpot system of Figure 2.11. This is a second-order system with one degree of freedom, so we must choose two state variables. Choosing

$$x_1 = x$$
$$x_2 = \dot{x} \tag{2.3}$$

the time rates of change of the two state variables are

$$\dot{x}_1 = x_2$$
$$\dot{x}_2 = -\frac{k}{m}x_1 - \frac{c}{m}x_2 + \frac{1}{m}F$$

These equations can be written in matrix notation

$$\dot{\mathbf{x}} = \mathbf{A}\mathbf{x} + \mathbf{B}\mathbf{u} \tag{2.4}$$

where

$$\mathbf{x} = \left[\begin{array}{c} x_1 \\ x_2 \end{array} \right]$$

$$\mathbf{A} = \left[\begin{array}{cc} 0 & 1 \\ -\dfrac{k}{m} & -\dfrac{c}{m} \end{array} \right]$$

$$\mathbf{u} = F$$

$$\mathbf{B} = \left[\begin{array}{c} 0 \\ \dfrac{1}{m} \end{array} \right]$$

Matrix **A** is called the *system matrix*. The system matrix is always square. Matrix **B** is the *input matrix*. The number of rows in the input matrix is the same as the number of state variables, and the number of columns is the same as the number of inputs. Equation (2.4) is the matrix form of the system state equation for a linear system. Note that the system and input matrices are not characteristic of the system. Different choices of state variables will result in different system and input matrices.

The state variables we defined may be considered internal states of a system. The state variables are not necessarily the system outputs. The outputs may consist of a subset of the states or may consist of a linear combination of the system states and the inputs. The output equation for a linear system is

$$\mathbf{y} = \mathbf{C}\mathbf{x} + \mathbf{D}\mathbf{u} \tag{2.5}$$

If we choose the output of the system depicted in Figure 2.11 to be the position of the mass (x), referring to Equation (2.3),

$$y = x_1$$

we have

$$\mathbf{C} = [\ 1 \quad 0\ ]$$

and

$$\mathbf{D} = 0$$

$\mathbf{C}$ is called the *output matrix*. $\mathbf{D}$ is called the *direct transmittance matrix* because if it is nonzero, the input is transmitted directly to the output.

The complete specification of a state-space block consists of the four matrices and the initial values of the state variables. The state-space block is frequently depicted as three blocks, as shown in Figure 2.16(a) (see, for example Kuo [3]), with the direct transmittance block ($\mathbf{D}$) omitted if there is no direct transmittance. The representation in Figure 2.16(b) is also common (see Middleton [5] and Examples 2.3 and 2.4), as is the depiction used by Simulink. Note that the state-space block solves the vector differential equation and, therefore, contains an integrator for each element of the state vector.

EXAMPLE 2.3   Electric hot water heater

Consider the electric water heater shown in Figure 2.17. We wish to model the heat loss to outside air as discussed in Middleton [5]. The heat capacity of the tank is C, and the ambient air temperature is T_0. Heat leaves the tank at the rate

$$heat\ out = k(T - T_0)$$

and enters (via the heater) at the rate

$$heat\ in = \frac{1}{CR}u^2$$

This is a single-degree-of-freedom first-order system, so only one state variable is required. We choose the state variable to be the temperature difference,

$$x = T - T_0$$

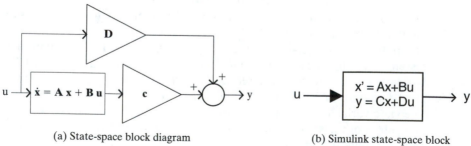

(a) State-space block diagram   (b) Simulink state-space block

FIGURE 2.16: State-space block

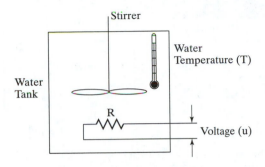

FIGURE 2.17: Electric hot water heater

and to avoid the nonlinearity due to u^2, we define

$$p = u^2$$

Thus,

$$\dot{x} = \frac{k}{C}x + \frac{1}{CR}p$$

The output is our state variable,

$$y = x$$

In this example, all the matrices are scalar; they have only one element:

$$\mathbf{A} = -\frac{k}{C}$$

$$\mathbf{B} = \frac{1}{CR}$$

$$\mathbf{C} = 1$$

$$\mathbf{D} = 0$$

EXAMPLE 2.4   State-space models

In this example, we will find state-space models for both sets of state variables defined in Example 2.2. Referring to Figure 2.15, we see that there is no input to this system. Therefore, we know that **B** and **D** may be ignored. Let's define the output to be the position of the right block (m_2). The force in the spring is

$$f_1 = k(x_2 - x_1)$$

and is positive acting on m_1 and negative acting on m_2. Therefore, for the first set of state variables

$$\dot{w}_1 = w_2$$

$$\dot{w}_2 = \frac{k}{m_1}(w_3 - w_1)$$

$$\dot{w}_3 = w_4$$

$$\dot{w}_4 = \frac{k}{m_2}(w_1 - w_3)$$

so

$$\mathbf{A} = \begin{bmatrix} 0 & 1 & 0 & 0 \\ -\dfrac{k}{m_1} & 0 & \dfrac{k}{m_1} & 0 \\ 0 & 0 & 0 & 1 \\ \dfrac{k}{m_2} & 0 & -\dfrac{k}{m_2} & 0 \end{bmatrix}$$

$$\mathbf{C} = \begin{bmatrix} 0 & 0 & 1 & 1 \end{bmatrix}$$

For the second set of state variables ($\mathbf{z}$), the force in the spring is dependent only on z_3. This makes the system matrix simpler. However, as the output is no longer a state variable, the output matrix must extract the position of the right block from z_1 and z_3:

$$\dot{z}_1 = z_2$$
$$\dot{z}_2 = \frac{k}{m_1}z_3$$
$$\dot{z}_3 = z_4$$
$$\dot{z}_4 = -\frac{k}{m_2}z_3$$

so

$$\mathbf{A} = \begin{bmatrix} 0 & 1 & 0 & 0 \\ 0 & 0 & \dfrac{k}{m_1} & 0 \\ 0 & 0 & 0 & 1 \\ 0 & 0 & -\dfrac{k}{m_2} & 0 \end{bmatrix}$$

The output is

$$\mathbf{y} = x_2 = z_1 + z_3$$

so the output matrix is

$$\mathbf{C} = \begin{bmatrix} 1 & 0 & 1 & 0 \end{bmatrix}$$

2.3.3 Multiple Input, Multiple Output Systems

The system state equation (2.4) for a linear system is not limited to a particular number of inputs, and the output equation (2.5) is not limited as to the number of outputs that may be defined. Many physical systems have numerous inputs and outputs, and the state-space block is a convenient means of representing such systems. Rather than show each input and output as a separate line, it is common to use vector lines to represent a set of signals, such as the input vector and the output vector. Vector lines are shown as thick lines or double lines. Depicted in Figure 2.18 are both common representations. Example 2.5 illustrates the development of a state-space model for a multiple input, multiple output system.

EXAMPLE 2.5   Three-mass system

Consider the system depicted in Figure 2.19. There are two inputs to the three-mass system. The outputs are the positions of each block. We'll choose as state variables the positions and

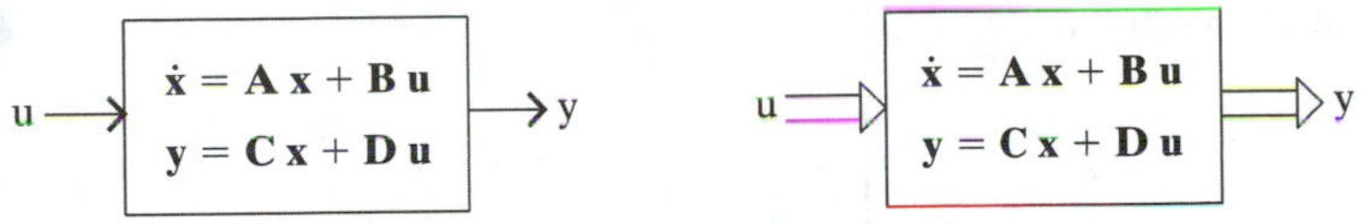

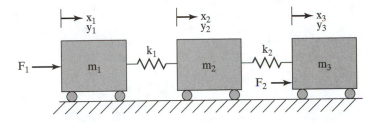

(a) Vector shown as heavy line

(b) Vector shown as double line

FIGURE 2.18: Multiple input, multiple output state-space block

FIGURE 2.19: Three-mass system with forcing functions

velocities of the blocks. The accelerations of the blocks are

$$\ddot{x}_1 = \frac{1}{m_1}[k_1(x_2 - x_1) + F_1]$$

$$\ddot{x}_2 = \frac{1}{m_2}[k_1(x_1 - x_2) + k_2(x_2 - x_2)]$$

$$\ddot{x}_3 = \frac{1}{m_3}[k_2(x_2 - x_3) + F_2]$$

The input vector is

$$\mathbf{u} = \begin{bmatrix} F_1 \\ F_2 \end{bmatrix}$$

We choose the following state variables:

$$z_1 = x_1$$
$$z_2 = \dot{x}_1$$
$$z_3 = x_2$$
$$z_4 = \dot{x}_2$$
$$z_5 = x_3$$
$$z_6 = \dot{x}_3$$

and the output variables are

$$z_1 = z_1$$
$$z_2 = z_3$$
$$z_3 = z_5$$

Thus, the system matrix is

$$
\mathbf{A} = \begin{bmatrix}
0 & 1 & 0 & 0 & 0 & 0 \\
-\dfrac{k_1}{m_1} & 0 & \dfrac{k_1}{m_1} & 0 & 0 & 0 \\
0 & 0 & 0 & 1 & 0 & 0 \\
\dfrac{k_1}{m_2} & 0 & -\dfrac{(k_1 + k_2)}{m_2} & 0 & \dfrac{k_2}{m_2} & 0 \\
0 & 0 & 0 & 0 & 0 & 1 \\
0 & 0 & \dfrac{k_2}{m_3} & 0 & -\dfrac{k_2}{m_3} & 0
\end{bmatrix}
$$

and the input matrix is

$$
\mathbf{B} = \begin{bmatrix}
0 & 0 \\
\dfrac{1}{m_1} & 0 \\
0 & 0 \\
0 & 0 \\
0 & 0 \\
0 & \dfrac{1}{m_3}
\end{bmatrix}
$$

The output matrix is

$$
\mathbf{C} = \begin{bmatrix}
1 & 0 & 0 & 0 & 0 & 0 \\
0 & 0 & 1 & 0 & 0 & 0 \\
0 & 0 & 0 & 0 & 1 & 0
\end{bmatrix}
$$

and the direct transmittance matrix is zero:

$$
\mathbf{D} = \begin{bmatrix}
0 & 0 \\
0 & 0 \\
0 & 0
\end{bmatrix}
$$

To complete the definition of the state-space model, it is necessary to specify the initial value of each element of the state vector.

2.4 DISCRETE SYSTEMS

A discrete system is a system that may be represented using difference equations and that operates on discrete signals. A discrete signal can be represented as a sequence of pulses, as shown in Figure 2.20.

A discrete system takes as its input one or more discrete signals and produces one or more discrete signals as its output. A discrete-time system is a discrete system that may change only at specific instants of time (Ogata [6]). In the majority of discrete-time control systems, the signals are not inherently discrete. In these systems, the discrete signals are extracted from continuous signals by a process known as *sampling*. Figure 2.21 illustrates the sampling process.

The type of sampling illustrated in Figure 2.21 is performed using two devices: a sampler and a zero-order hold. The sampler periodically closes a switch for an instant. Each time the switch is momentarily closed, a pulse of very short (theoretically zero) duration, and equal in magnitude to the input signal, is produced [Figure 2.21(b)]. The spacing between the pulses is the sampling period. The zero-order hold follows the sampler and clamps its output at the last value of its input, producing a stair-step

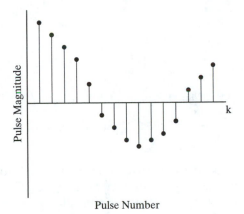

FIGURE 2.20: Discrete signal

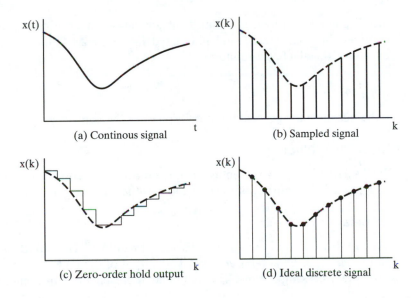

FIGURE 2.21: Sampling a continuous signal to produce a discrete signal

signal, as shown in Figure 2.21(c). A discrete controller requires a sequence of numbers as its input. The analog to digital (A/D) converter is a device that converts the stair-step signal of Figure 2.21(c) into a sequence of numbers, as represented by the pulse sequence shown in Figure 2.21(d). The combination of the sampler and zero-order hold block is illustrated in Figure 2.22, along with the Simulink equivalent.

We refer to signals that vary with time using the notation $x(t)$. The corresponding notation for discrete signals is $x(k)$, where k is the pulse number. The mapping from continuous time to sample space is

$$x(k) = x(kT)$$

where T is the sampling period.

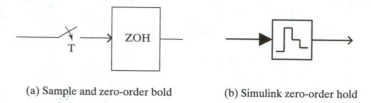

(a) Sample and zero-order bold   (b) Simulink zero-order hold

FIGURE 2.22: Sample and zero-order hold

We will now discuss the discrete counterparts to the continuous system block diagram components discussed in Section 2.2. These discrete components may be physically realized as digital circuit elements or as computer instructions in microprocessor-based controllers.

2.4.1 Discrete Gain Block

The discrete gain block is identical in appearance to the continuous gain block described in Section 2.2.1 and illustrated in Figure 2.1. The discrete gain block represents the algebraic equation

$$y(k) = ax(k)$$

so the output is the present input multiplied by the gain.

2.4.2 Discrete Sum Block

The discrete sum block is identical in appearance to the continuous sum block shown in Figure 2.3. The output of the sum block is the algebraic sum of its inputs.

2.4.3 Unit Delay

The *unit delay* is the fundamental discrete-time block. The unit delay is also called a *shift register* or *time-delay element*. The output of the unit delay block is the input at the previous sample time. The unit delay represents the difference equation

$$y(k) = x(k - 1)$$

In Figure 2.23, two common representations of the unit delay and the Simulink equivalent are shown.

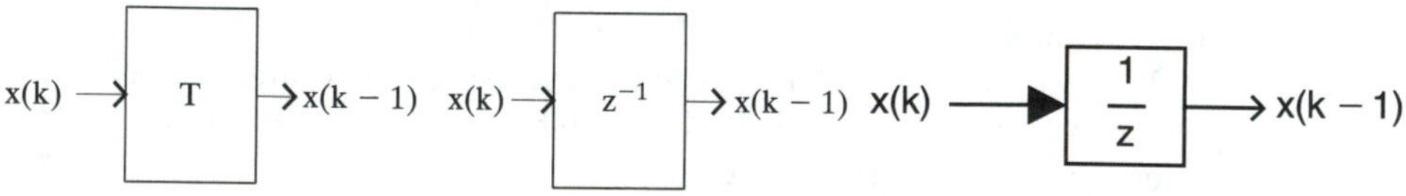

$$x(k) \longrightarrow \boxed{\quad T \quad} \longrightarrow x(k-1) \qquad x(k) \longrightarrow \boxed{\quad z^{-1} \quad} \longrightarrow x(k-1) \qquad x(k) \longrightarrow \boxed{\dfrac{1}{z}} \longrightarrow x(k-1)$$

(a) Unit delay block   (b) Z transform notation   (c) Simulink unit delay

FIGURE 2.23: Unit delay block

2.4.4 Discrete-Time Integrator Block

The output of the discrete-time integrator block is an approximation of the time integral of the input signal. That is, it is a discrete approximation of a continuous integrator. Thus, the output of the discrete-time integrator block approximates

$$y(k) = y(k-1) + \int_{T(k-1)}^{Tk} u(t)\, dt$$

where $u(t)$ is the input to the integrator, $y(k)$ is the output, and T is the sample period.

Three standard implementations of the discrete-time integrator block are *forward Euler integration* (also known as forward rectangular integration), *backward Euler integration* (backward rectangular integration), and *trapezoidal integration*. All three implementations approximate the area under a curve as the sum of the areas of a finite number of rectangles.

Note that you will sometimes see the definitions of forward and backward Euler integration reversed from the Simulink convention, which is used here. For example, Kuo [3] referred to the method Simulink calls forward Euler integration as backward rectangular integration, and the method Simulink calls backward Euler integration as forward rectangular integration.

Forward Euler. Forward Euler integration is based on the approximation $u(t) = u(T(k-1))$. Thus, forward Euler integration approximates

$$y(k) = y(k-1) + Tu(T(k-1)) \tag{2.6}$$

Taking the Z transform of Equation (2.6),

$$Y(z) = z^{-1}Y(z) + Tz^{-1}U(z)$$

Rearranging terms, the Z transfer function of the forward Euler integrator is

$$\frac{Y(z)}{U(z)} = \frac{T}{z-1}$$

This is graphically depicted in Figure 2.24.

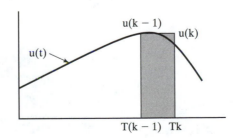

FIGURE 2.24: Forward Euler integration

Backward Euler.   Backward Euler integration is based on the approximation $u(t) = u(Tk)$. Thus, backward Euler integration approximates

$$y(k) = y(k-1) + Tu(T(k))$$

The corresponding Z transfer function is

$$\frac{Y(z)}{U(z)} = \frac{Tz}{z-1}$$

This is graphically depicted in Figure 2.25.

Trapezoidal Integration.   Trapezoidal integration is based on the approximation

$$u(t) = \frac{u(Tk) + u(T(k-1))}{2}$$

Thus, trapezoidal integration approximates

$$y(k) = y(k-1) + \frac{Tu(T(k)) + Tu(T(k-1))}{2}$$

The corresponding Z transfer function is

$$\frac{Y(z)}{U(z)} = \frac{T(z+1)}{2(z-1)}$$

The rule for trapezoidal integration is illustrated in Figure 2.26.

Discrete-Time Integrator Blocks.   Shown in Figure 2.27 are representations of forward Euler, backward Euler, and trapezoidal integrator blocks. The Simulink Discrete-Time Integrator block can be set to use any of these methods. Forward and backward rectangular integration are probably equally common in practice; trapezoidal is somewhat less common. All three produce approximately the same results when used properly. As with continuous integrators, the complete system specification includes the initial value of the integrator output.

2.4.5 Simple Discrete System Models

The discrete blocks we've described so far may be used to build a model of any linear discrete system. In Example 2.6, the development of a block diagram for a simple economic problem is illustrated.

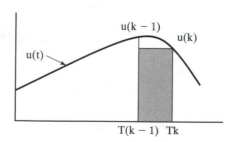

FIGURE 2.25: Backward Euler integration

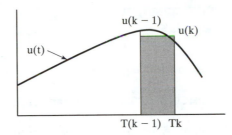

FIGURE 2.26: Trapezoidal integration

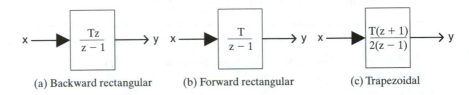

(a) Backward rectangular   (b) Forward rectangular   (c) Trapezoidal

FIGURE 2.27: Discrete integrator blocks

EXAMPLE 2.6   Loan amortization

In this example, we will model the amortization of an automobile loan. At the end of each month, the loan balance $b(k)$ is the sum of the balance at the beginning of the month $[b(k-1)]$ and the interest for the month $[ib(k)]$, less the end-of-month payment $p(k)$. Thus, the balance at the end of month k is

$$b(k) = rb(k-1) - p(k)$$

where $r = i + 1$, and i is the monthly interest rate.

The block diagram in Figure 2.28 will model this system. Note that the complete system specification includes the initial loan balance.

2.4.6   Discrete Transfer Function Block

The discrete transfer function is analogous to the continuous transfer function we discussed in Section 2.2.6. The discrete transfer function is the ratio of the Z

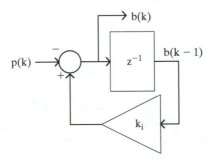

FIGURE 2.28: Block diagram of loan amortization

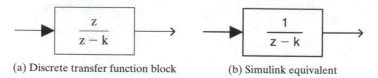

(a) Discrete transfer function block   (b) Simulink equivalent

FIGURE 2.29: Discrete transfer function block

transform of the input to a system or subsystem to the Z transform of the output. Like the continuous transfer function, the discrete transfer function provides a compact notation for the input–output relationship of a system. We'll illustrate the definition with a simple example (Example 2.7).

EXAMPLE 2.7   Loan amortization using discrete transfer function

Consider the loan amortization problem discussed in Example 2.6. The difference equation describing the loan amortization is

$$b(k) = k_i b(k-1) - p(k)$$

Taking the Z transform,

$$B(z) = k_i z^{-1} B(z) - P(z)$$

Rearranging terms, we get the transfer function:

$$\frac{B(z)}{P(z)} = \frac{z}{z - k_i}$$

As with the continuous transfer function, the denominator of the transfer function is the characteristic equation of the system. The roots of the characteristic equation are the eigenvalues of the system. The discrete transfer function block is depicted in Figure 2.29, along with its Simulink equivalent.

2.5 DISCRETE STATE-SPACE BLOCK

The state-space concept discussed in Section 2.3 for continuous systems is also useful for modeling discrete systems. Whereas the state variables in continuous systems represent derivatives or algebraic combinations of derivatives, the state variables in discrete systems represent portions of sequences or algebraic combinations of portions of sequences. Thus, the discrete equivalent to Equation (2.1) is the general form of the discrete state-space model of a dynamical system:

$$\mathbf{x}(k+1) = \mathbf{f}(\mathbf{x}, \mathbf{u}, k)$$

where, as before, $\mathbf{x}$ is the state vector and $\mathbf{u}$ is the input vector. The corresponding output equation is

$$\mathbf{y}(k) = \mathbf{g}(\mathbf{x}, \mathbf{u}, k)$$

The matrix notation for linear discrete systems is similar to the model for continuous systems. The same four matrices (system matrix, input matrix, output matrix, and direct transmittance matrix) are used. Thus the system equation is

$$\mathbf{x}(k+1) = \mathbf{A}\mathbf{x}(k) = \mathbf{B}\mathbf{u}(k)$$

The output equation is

$$y(k) = \mathbf{C}x(k) + \mathbf{D}u(k)$$

Illustrated in Example 2.8 is the definition of a linear state-space model. In Example 2.9, a more complex model is illustrated.

EXAMPLE 2.8   Discrete state-space formulation

Consider a system that can be described by the difference equation

$$y(k + 1) = y(k) - 2y(k - 1) + u(k)$$

Define state variables

$$x_1(k) = y(k - 1)$$

$$x_2(k) = y(k)$$

The state equations are

$$x_1(k + 1) = x_2(k)$$

$$x_2(k + 1) = x_2(k) - 2x_1(k) + u(k)$$

In matrix form, we have

$$\mathbf{x}(k + 1) = \begin{bmatrix} 0 & 1 \\ -2 & 1 \end{bmatrix} \mathbf{x}(k) + \begin{bmatrix} 0 \\ 1 \end{bmatrix} \mathbf{u}$$

so the system matrix is

$$\mathbf{A} = \begin{bmatrix} 0 & 1 \\ -2 & 1 \end{bmatrix}$$

and the input matrix is

$$\mathbf{B} = \begin{bmatrix} 0 \\ 1 \end{bmatrix}$$

There is no direct transmittance, so $\mathbf{D}$ is zero. The output variable is $y(k)$, so

$$\mathbf{C} = \begin{bmatrix} 0 & 1 \end{bmatrix}$$

EXAMPLE 2.9   More complex discrete state-space formulation

Suppose a linear system can be described by the difference equation

$$y(k + 1) + 5y(k) + 3y(k - 1) = u(k) + 2u(k - 1)$$

where $y(k)$ is the output. We choose the following state variables to make $u(k - 1)$ implicit:

$$x_1(k) = y(k - 1) - \frac{2}{3}u(k - 1)$$

$$x_2(k) = y(k)$$

The state equations are

$$x_1(k + 1) = x_2(k) - \frac{2}{3}u(k)$$

$$x_2(k + 1) = -3x_1(k) - 5x_2(k) + u(k)$$

The system matrix is

$$A = \begin{bmatrix} 0 & 1 \\ -3 & -5 \end{bmatrix}$$

and the input matrix is

$$B = \begin{bmatrix} -\dfrac{2}{3} \\ 1 \end{bmatrix}$$

The output matrix is

$$C = \begin{bmatrix} 0 & 1 \end{bmatrix}$$

and the direct transmittance matrix is zero:

$$D = \begin{bmatrix} 0 \end{bmatrix}$$

2.6 NONLINEAR BLOCKS

The block diagram concept can be generalized to include a wide variety of blocks in addition to the linear systems blocks we've defined so far. Normally, a block is labeled with text or a picture that makes it easy to understand the purpose of the block. Some of the more common blocks are listed below, although this list is far from complete.

- Product—product of two scalar input signals
- Abs—absolute value of the input signal
- Logical operators (AND, OR, NOT)
- Relational operators (>, <)
- Sign—signum nonlinearity
- Saturation function
- Transport delay
- Table lookup (interpolate in a table)

2.7 HYBRID BLOCK DIAGRAMS

A *hybrid system* contains discrete and continuous components. The most common hybrid system consists of a continuous physical process that is controlled using discrete logic components or a microprocessor. Illustrated in Figure 2.30 is a hypothetical hybrid system adapted from Ogata [6], in which the controller includes continuous and discrete blocks. We'll describe each block in Figure 2.30. Referring to the letters (a, b, etc.) in the blocks, we'll proceed clockwise, starting with the continuous process block.

a. The continuous process could be any physical process, such as a chemical plant tank heater, an aircraft autopilot, or a robot arm. We are interested in controlling a particular parameter of the process, such as the temperature of the tank contents, the aircraft pitch attitude, or the angle of one of the robot arm rotary joints.

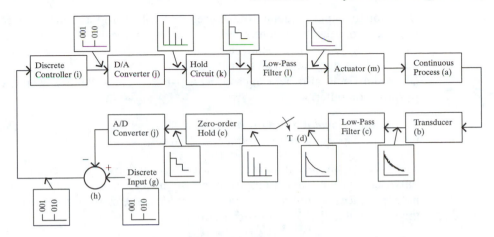

FIGURE 2.30: Hybrid system

b. The transducer is a device that maps each possible value of the parameter we're controlling to a unique value of the output signal. For example, a temperature transducer may consist of a thermocouple that produces a voltage proportional to the temperature. The aircraft pitch transducer may produce a voltage proportional to a gyroscope gimbal angle, and the robot joint angle transducer might consist of a voltage signal produced by a potentiometer.

c. Transducers frequently produce output signals that contain noise. To reduce the noise from the signal before the signal is sampled, it is common to follow the transducer with a low-pass filter. In this application, the low-pass filter is frequently called an anti-aliasing filter. The block diagram representation of the low-pass filter is usually a transfer function block, although the filter could be described using the primitive blocks described in Section 2.2 or a continuous state-space block as described in Section 2.3.

d. The sampler is the first step in producing the discrete signal. It produces a sequence of pulses, as described in Section 2.4.

e. The zero-order hold clamps the sampler output to create the stair-step signal. Up to this point, the signal is a physical quantity, such as voltage or current.

f. The A/D converter produces a sequence of numbers that corresponds to the stair-step input signal. Frequently, the zero-order hold and A/D converter are combined in a single device.

g. The discrete input is the desired numerical value (the set point) of the parameter being controlled. This could be a value input at a computer console, or it could be a value set by positioning a bank of switches.

h. The discrete summer computes the difference in the desired value of the parameter and the measured value. This difference is the error in the controlled parameter, and it is the input to the discrete controller.

i. Based on the input sequence, the discrete controller produces an output sequence intended to cause the controlled parameter to reach the desired value (g). The discrete controller could be built using the primitive blocks described in Section 2.4, a discrete transfer function block, or a discrete state-space block. The controller could also be a microprocessor-based controller, or even a general-purpose computer.

j. The *digital to analog (D/A) converter* transforms an input sequence of numbers into a corresponding output sequence of pulses.

k. The hold circuit, usually a zero-order hold, transforms the discontinuous input pulse sequence into a continuous output sequence. In this example, a zero-order hold produces a stair-step signal. In practice, the zero-order hold is usually incorporated in the D/A converter device.

l. A low-pass filter smooths the stair-step signal. It is frequently necessary to smooth this signal to prevent damage to physical components or to conserve the energy consumed by the actuator. For example, a series of step inputs to an aircraft autopilot could produce a rough ride for the passengers.

m. The actuator is a device that changes a characteristic of the controlled process. For the tank heater, the actuator could be a steam valve. The aircraft autopilot would control the elevator position. The robot joint control might use a variable voltage regulator that controls drive motor torque.

An endless number of variations of hybrid controllers is possible. The controller can be simplified considerably if the transducer is a digital device. For example, pulse encoders are a popular means of sensing the position of a rotary joint.

2.8   SUMMARY

In this chapter, we discussed the most common block diagram elements for continuous and discrete systems. We discussed the construction of simple block diagrams of linear systems using primitive blocks, transfer function blocks, and state-space blocks. We briefly mentioned the many possible nonlinear blocks. Finally, we discussed hybrid systems and looked at the block diagram of a hypothetical hybrid system.

REFERENCES

[1] Dorf, Richard C., *Modern Control Systems*, 7th ed. (Reading, Mass.: Addison-Wesley Publishing Company, Inc., 1995). This introductory text in control systems uses MATLAB extensively. Numerous examples of models of dynamical systems are presented in the book, many including block diagrams.

[2] Harman, Thomas L., Dabney, James, and Richert, Norman, *Advanced Engineering Mathematics with MATLAB* (Pacific Grove, Ca.: Brooks/Cole Publishing Company, 2000). This book presents a practical approach to advanced engineering mathematics. MATLAB is used in many examples and in the extensive set of worked problems at the end of each chapter.

[3] Kuo, Benjamin C., *Automatic Control Systems*, 7th ed. (Englewood Cliffs, N.J.: Prentice Hall, 1995), 274. An excellent basic text on control systems.

[4] Lewis, Paul H., and Yang, Charles, *Basic Control Systems Engineering* (Upper Saddle River, N.J.: Prentice Hall, 1997). This text provides an introduction to control systems analysis and design and includes a brief introduction to Simulink. Simulink is used in many of the examples.

[5] Middleton, Richard H., and Goodwin, Graham C., *Digital Control and Estimation: A Unified Approach* (Englewood Cliffs, N.J.: Prentice Hall, 1990), 13–14. This is an advanced text on digital control systems. It provides a balanced mixture of theory and applications.

[6] Ogata, Katsuhiko, *Discrete-Time Control Systems*, 2nd ed. (Englewood Cliffs, N.J.: Prentice Hall, 1994), 5–7. This is an excellent text on discrete control systems. Many examples are presented with MATLAB solutions, and there are many block diagrams that can be easily converted into Simulink models. This text is very readable and would be particularly useful for self-study or as a reference.

[7] Phillips, Charles L., and Nagle, H. Troy, *Digital Control System Analysis and Design*, 3rd ed. (Englewood Cliffs, N.J.: Prentice Hall, 1995). This book provides a comprehensive coverage of discrete-time control systems analysis and design. It presents many practical examples and makes good use of MATLAB.

3

Quick Start

In this chapter, we will discuss the basics of building and executing Simulink models. We will start with a simple first-order system. Next, we will build a more complex model that includes feedback and illustrates several important procedures in Simulink programming. Finally, we will discuss the Simulink Help system.

3.1 INTRODUCTION

Simulink is a powerful programming language, and a big part of that power is its ease of use. In this chapter, we will introduce Simulink programming by building and executing two simple models. Our purpose here is to cover the basics of model building and execution. We will discuss these topics in more detail in later chapters.

The examples shown in this chapter (and the rest of the book) were produced using Simulink 5.0 and MATLAB 6.5 in the Microsoft Windows XP environment. There are very few differences between using Simulink in the Windows XP environment and using it in the X-Windows environment; we will point out those differences where necessary.

3.1.1 Typographical Conventions

Before we build the first model, we need to establish some typographical conventions:

Computer Type.   All computer input, output, variable names, and command names are shown in `monospaced sans-serif` font.

Using Menus.   The Simulink user interface employs pull-down menus located on a menu bar at the top of the Simulink model window. To facilitate discussion of the various menu choices, we will use the convention: **Menu bar choice:Pull-down menu choice**. For example, choosing "File" followed by "Save As" (as shown in Figure 3.1) will be written **File:Save As**.

Dialog Box Fields.   Simulink makes extensive use of dialog boxes to set simulation parameters and to configure blocks. Dialog box fields are indicated in **bold type**. For example, choosing **File:Save As** opens the dialog box shown in Figure 3.2. The fields in this dialog box are **Save in**, **File name**, and **Save as type**.

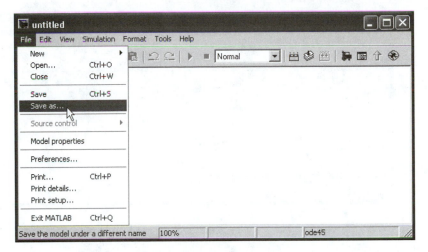

FIGURE 3.1: Menu selection

FIGURE 3.2: **File:Save As** dialog box

3.2   BUILDING A SIMPLE MODEL

Let's start by building a Simulink model that solves the differential equation

$$\dot{x} = \sin(t) \tag{3.1}$$

with initial condition $x(0) = 0$.

Simulink is an extension of MATLAB, and it must be invoked from within MATLAB. Start Simulink by clicking the Simulink icon on the MATLAB toolbar (Microsoft Windows), as shown in Figure 3.3, or by entering the command `Simulink` at the MATLAB prompt. On an X-Windows system, enter the command `Simulink` at the MATLAB prompt.

The Simulink library browser, shown in Figure 3.4, will open. Click the New Window icon, opening an empty model window, as shown in Figure 3.5. It is this empty model window, initially named `untitled`, in which you will build the Simulink model.

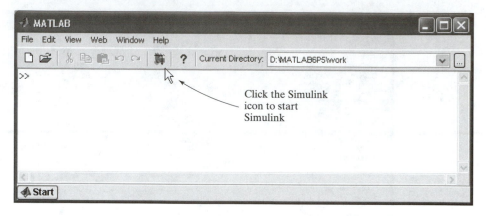

FIGURE 3.3: Starting Simulink

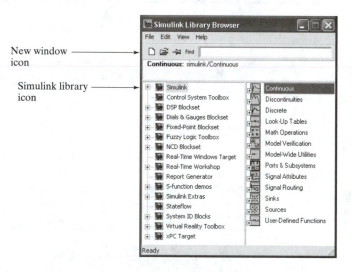

FIGURE 3.4: Simulink Library Browser window

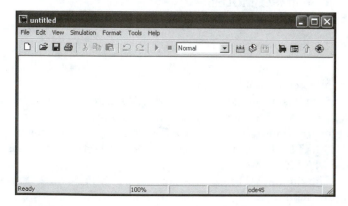

FIGURE 3.5: Empty model window

Double-click the Simulink library icon, or click the +, which expands the Simulink library as shown. Notice that the right pane contains a list of sublibraries.

Select the Sources library to open the library in the right pane.

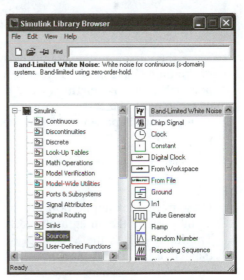

Drag the Sine Wave block from the Sources block library to the model window, positioning it as shown. A copy of the block is placed in the model window. Note that, in the figure, we resized the model window to save space.

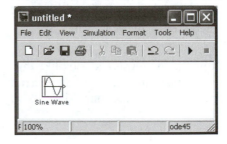

Open the Continuous block library, and drag an Integrator block to the model window.

Open the Sinks block library, and drag a Scope block to the model window.

Next, connect the blocks with signal lines to complete the model.

Place the cursor on the output port of the Sine Wave block. The output port is the > symbol on the right edge of the block. The cursor changes to a cross-hair shape when it's on the output port.

Drag from the output port to the input port of the Integrator block. The input port is the > symbol on the left edge of the block. As you drag, the cursor retains the cross-hair shape. When the cursor is on the input port, it changes to a double-lined cross-hair, as shown.

Now the model should look like this. The signal line has an arrowhead indicating the direction of signal flow.

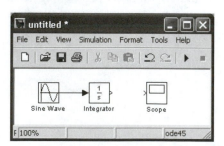

Draw another signal line from the
Integrator output port to the Scope
input port, completing the model.

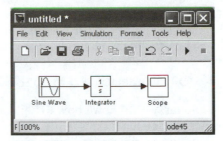

Double-click the Scope block, opening a Scope window, as shown in Figure 3.6.

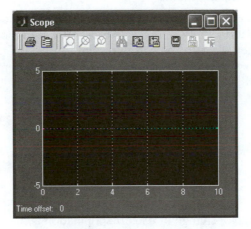

FIGURE 3.6: Scope window

Choose **Simulation:Start** from the menu bar in the model window. The simulation will execute, resulting in the scope display shown in Figure 3.7.

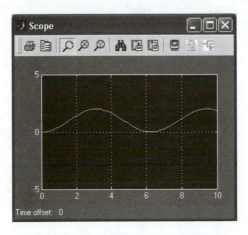

FIGURE 3.7: Scope display after executing the model

To verify that the plot shown in Figure 3.7 represents the solution to Equation (3.1), you can solve Equation (3.1) analytically, with the result $x(t) = 1 - \cos t$. Thus, the Simulink model solved the differential equation correctly.

3.3   A MORE COMPLICATED MODEL

So far, we've shown how to build a simple model. There are a number of additional model-building skills that you'll need to acquire. In this section, we use a model of a biological process to illustrate several additional skills: branching from signal lines, routing signal lines in segments, flipping blocks, configuring blocks, and configuring the simulation parameters.

Scheinerman [1] described a simple model of bacteria growth in a jar. Assume that the bacteria are born at a rate proportional to the number of bacteria present, and that they die at a rate proportional to the square of the number of bacteria present. If x represents the number of bacteria present, then the bacteria are born at the rate

$$\text{birth rate} = bx$$

and die at the rate

$$\text{death rate} = px^2$$

The total rate of change of bacteria population is the difference between birth rate and death rate. This system can therefore be described with the differential equation

$$\dot{x} = bx - px^2 \tag{3.2}$$

Let's build a model of this dynamical system assuming that $b = 1/\text{hour}$ and $p = 0.5/\text{bacteria-hour}$. Then, we'll compute the number of bacteria in the jar after 1 hour, assuming that initially there are 100 bacteria present.

Open a new model window as discussed earlier.

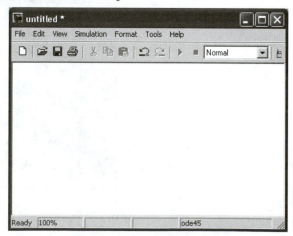

This is a first-order system, so one integrator is required to solve the differential equation. The input to the integrator is $\dot{x}$, and the output is x. Open the Continuous block library and drag the integrator block to the position shown.

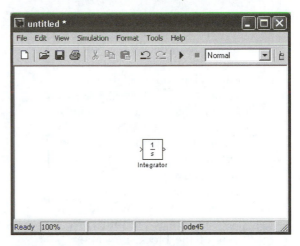

Drag two Gain blocks from the Math block library, and place them as shown. Notice that the name of the second Gain block is Gain1. Simulink requires each block to have a unique name. We'll discuss changing the name in Chapter 4.

Drag a Sum block and a Product block from the Math block library, and position them as shown. We will use the Product block to compute x^2.

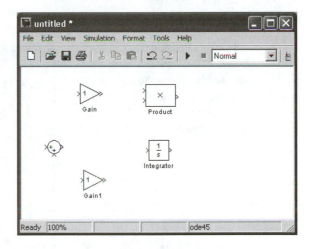

Open the Sinks block library and drag a Scope block to the model window, as shown.

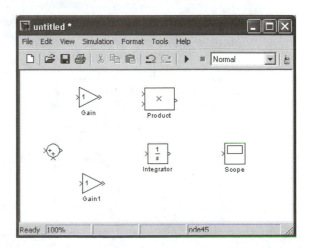

The default orientation of all the blocks places input ports on the left edge of the block and output ports on the right edge. The model will be much easier to read if we flip the Product block and the Gain blocks so that the input ports are on the right edge, and the output ports are on the left edge. Starting with the Product block, click the block once to select it. Notice the handles that appear at the four corners of the block, indicating that it is selected.

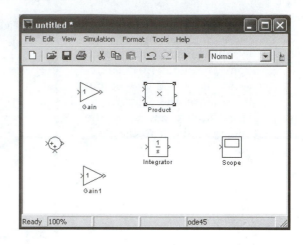

Choose **Format:Flip Block** from the model window menu bar. Now the inputs are on the right, and the output is on the left. Repeat the flipping operation for each Gain block.

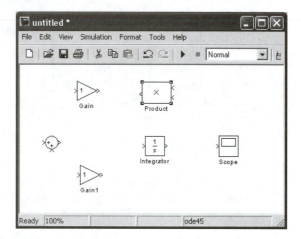

Draw a signal line from the output of the Sum block to the input of the Integrator block, and another from the output of the Integrator to the input of the Scope block.

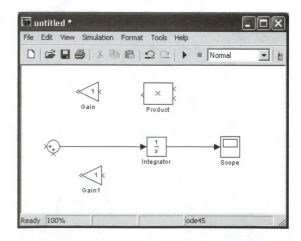

Next, we need to branch from the signal line connecting the Integrator and Scope blocks to feed the value of x to the lower Gain block. Press and hold the Control key and click the signal line. The cursor will change to a cross-hair shape. Continue to depress the mouse button and release the key.

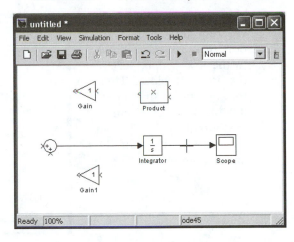

If the mouse has two or three buttons, clicking and dragging using the right mouse button is an alternative to branching using the Control key.

Drag directly to the input port of the Gain block. Notice that the signal line is dashed, and the cursor changes to a double cross-hair when it is on the input port of the gain block. Simulink automatically routes the signal line using 90-degree bends.

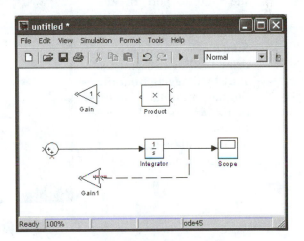

In a similar manner, branch from the signal line connecting the Integrator and Scope blocks to the top input port of the Product block.

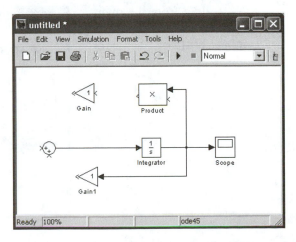

Branch from the signal line entering the upper input port of the Product block to the lower input port of the Product block. Thus, the output of the Product block is x^2. Connect the output of the Product block to the input of the upper Gain block.

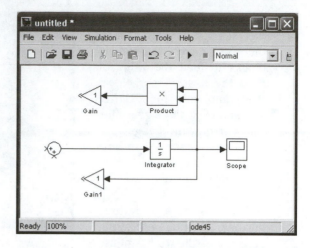

Draw the signal line from the output port of the upper Gain block to the upper input port of the Sum block in segments. To draw the line in segments, start by dragging from the output port to the location of the first bend. Release the mouse button. The signal line will be terminated with an open arrowhead.

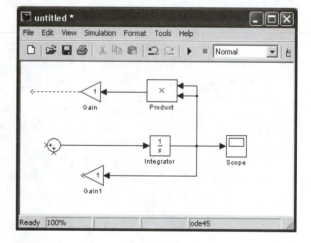

Next, drag from the open arrowhead to the upper input port of the Sum block.

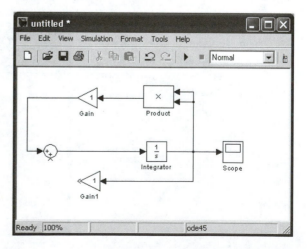

In a similar manner, draw the signal line from the output of the lower Gain block to the lower input of the Sum block.

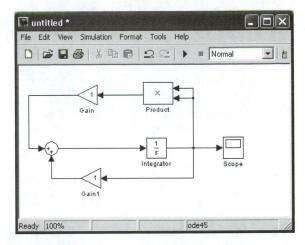

The model is now complete, but several of the blocks must be configured. Currently, the value of gain for both Gain blocks is the default value of 1.0. The Sum block adds its two inputs instead of computing the difference. Finally, the initial value of the integrator output [the initial number of bacteria, x] must be set, as it defaults to 0. Start with the Gain blocks.

Double-click the upper Gain block, which corresponds to coefficient p. The Gain block dialog box will be displayed. Change the default value in field **Gain** to 0.5. Click **OK**.

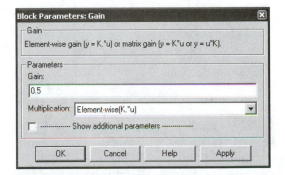

Notice that the value of gain on the block icon is now 0.5.

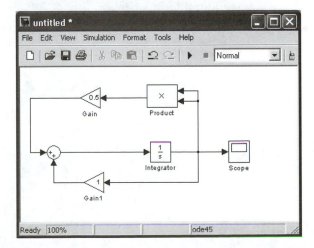

Double-click the Sum block, opening the Sum block dialog box. Change the first (+) sign in List of signs to (−) (a minus sign) as shown. Thus, List of signs should contain (| − +). Click **OK**.

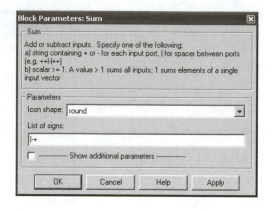

Notice that the sign on the higher Sum block input port changed to (−).

Now the Sum block is configured to compute the value of $\dot{x}$ according to Equation (3.2), after substituting in the values of b and p.

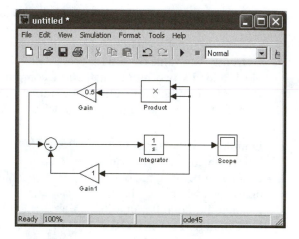

The final configuration task is to set the initial value of number of bacteria (x). Double-click the Integrator block, opening the Integrator dialog box. Set **Initial condition** to 100. Click **OK**.

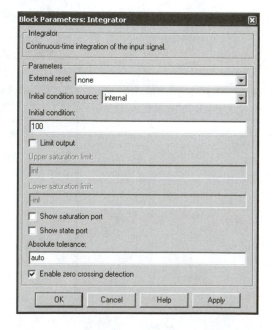

The simulation start time defaults to 0, and the stop time defaults to 10.0. To change the stop time to 1, open the Simulation parameters dialog box by choosing **Simulation:Parameters** from the model window menu bar. Set **Stop time** to 1, then click **OK**.

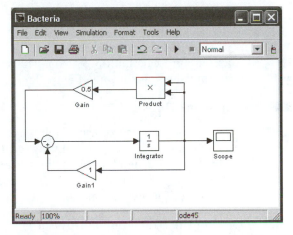

The model is now complete and ready to run. It's always a good idea to save a model before running it. To save the model, choose **File:Save** from the model window menu bar (or click the ▤ icon), and enter a file name, say Bacteria, without an extension. Simulink will save the model with the extension .mdl and change the model window name from untitled to the name you entered.

Open the Scope by double-clicking the Scope block. Then choose **Simulation:Start** to run the simulation. The scope display will be as shown in Figure 3.8.

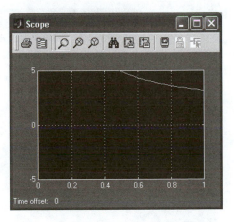

FIGURE 3.8: Scope display after running the bacteria growth model

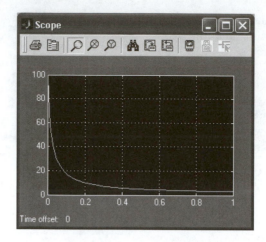

FIGURE 3.9: Scope display after running the bacteria growth model

Click on autoscale button 🔍 icon. The scope will resize the scale to fit the entire range of values, as shown in Figure 3.9.

3.4   THE Simulink HELP SYSTEM

Simulink includes an extensive Help system in the form of an online HTML format User's Guide and an integrated browser. Detailed online documentation for all of the blocks in the Simulink block library is available through the browser, shown in Figure 3.10. Additionally, online help is available by clicking the **Help** button in the dialog boxes for **Simulation:Parameters** and **Edit:Block Properties**.

3.4.1   Opening the User's Guide Browser

The procedure for opening the user's guide browser for a particular block is as follows:

Double click a block,

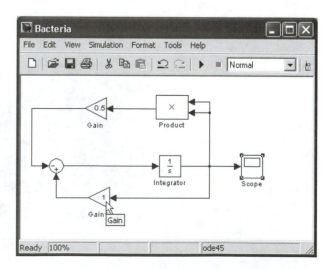

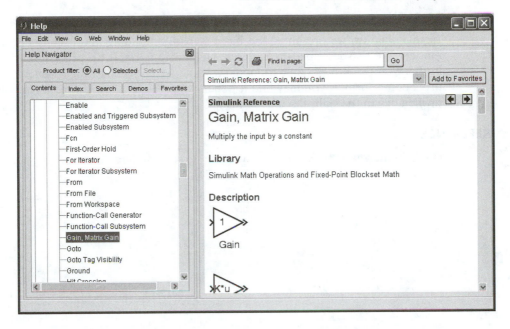

FIGURE 3.10: Simulink user's guide browser

opening the block dialog box. Click
the **Help** button.

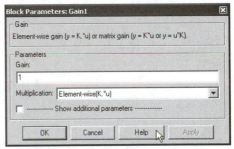

The browser (Figure 3.10) will open to the page for the block you selected.

3.4.2   User's Guide Browser Window

There are two frames to the user's guide browser window. The left frame contains
navigation information, and the right frame contains the selected text and graphics.
The four tabs on the navigation frame allow you to display a table of contents, an
alphabetical index, a text search window, and a window containing a customizable
list of favorite links. The contents tab is shown in Figure 3.10. Each of the tabbed
lists operates in a manner similar to the Simulink Library Browser. You can add any
page to the list of favorites by choosing the **Add to Favorites** button while the page
is displayed in the right frame. The browser window is actually a special-purpose
html browser, and it can also display Web pages (such as www.mathworks.com) if
the computer is connected to the Internet. In fact, the default entry in the favorites
list is for the MathWorks technical support Web page.

3.5 SUMMARY

In this chapter, we presented the basic steps for building and executing Simulink models. We also described the Simulink Help system. Using the procedures discussed in this chapter, it is possible to model a wide variety of dynamical systems. Before proceeding further, you may find it beneficial to build a few simple models to gain proficiency in the skills discussed thus far.

REFERENCE

[1] Scheinerman, Edward C., *Invitation to Dynamical Systems*. (Upper Saddle River, N.J.: Prentice Hall, 1996), 22–24.

4

Model Building

In this chapter, we will explain the mechanics of model building in detail. The procedures discussed here will enable you to build models that are easy to interpret. You will also learn how to select and configure a differential equation solver and how to print a Simulink model and embed a model in a word-processing document.

4.1 INTRODUCTION

In Chapter 3 we discussed the basics of building and running a Simulink model. Using the procedures discussed there and on the online Help system, you can build extremely complex models. However, as models get more complex, they become more difficult to interpret. The procedures discussed in this chapter will enable you to make your models easier to understand. First, careful arrangement of blocks and signal lines can make the relationships easier to follow. Next, naming blocks and signal lines and adding annotations to the model can make the purpose of the model elements easier to understand.

We will also describe the **Simulation:Parameters** dialog box that provides extensive options for selecting and configuring the differential equation solver used to perform model simulation. In addition to selecting a solver most appropriate for your problem, you can control the spacing of output points, the generation of error and warning messages, and even send internal simulation data to the MATLAB workspace.

Finally, we will explain how to print your models. You can print directly to a printer or embed an image of the model in a word-processing document.

Our purpose in this chapter is to explain the mechanics of model building: manipulating blocks, drawing and editing signal lines, annotating the model, and so on. Once you master these mechanics, you will be ready to begin building models using the procedures covered in the subsequent chapters.

4.1.1 Elements of a Model

A Simulink model consists of three types of elements: *sources*, the *system* being modeled, and *sinks*. Figure 4.1 illustrates the relationship among the three elements. The central element, the system, is the Simulink representation of a block diagram of the dynamical system being modeled. The sources are the inputs to the dynamical system. Sources include constants, function generators such as sine waves and step functions, and custom signals you create in MATLAB. Source blocks are found in the Sources block library. The output of the system is received by sinks. Examples

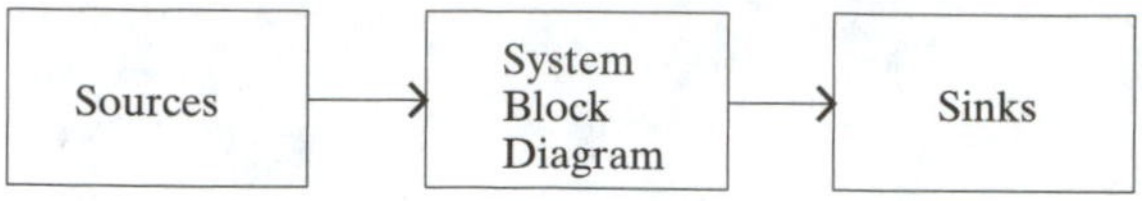

FIGURE 4.1: Elements of a Simulink model

of sinks are graphs, oscilloscopes, and output files. Sink blocks are found in the Sinks block library.

Frequently, Simulink models lack one or more of these three elements. For example, you might wish to model the unforced behavior of a system initially displaced from its equilibrium state. Such a model would have no inputs, but it would have system blocks (Gain blocks, Integrators, etc.) and probably sinks. It is also possible to build a model that has sources and sinks, but no system blocks. Suppose that you need a special signal composed of the sum of several functions. You could easily generate the signal using Simulink source blocks, and send the signal to the MATLAB workspace or to a disk file.

4.2   OPENING A MODEL

In Chapter 3, we discussed creating a new model by clicking the new window icon from the Simulink toolbar. We also discussed saving a model using **File:Save**. To use an existing model, click the open file icon from the menu bar of the Simulink window and select the file.

The default directory will be the current default directory in the MATLAB session from which you started Simulink. If you change to a different directory using the Simulink file menu, the default directory will not change. It will remain the same as the current directory in the MATLAB session, so if you wish to open another model in the same directory as the model you previously opened, and that model was not in the directory currently active in the MATLAB session, you will have to navigate once again to the appropriate directory. Therefore, it is frequently convenient to change to the directory containing the Simulink model in the MATLAB session (using the cd command), then open the model using the Simulink file open button.

Although it is not necessary, we recommend that you set up a directory for your Simulink models that is separate from the MATLAB directory structure. This will make it easier to keep track of your files. It will also eliminate the possibility of over-writing or deleting your files if you upgrade MATLAB and Simulink in the future.

An alternative method for opening an existing model is to type the name of the model as a command at the MATLAB prompt. MATLAB will search for the model via the MATLAB path, starting in the current directory. For example, if the model is named examp.mdl, enter the command examp at the MATLAB prompt.

4.3   MODEL WINDOW

You have several options for customizing the Simulink model window to suit your personal preferences. The model window for a simple system is shown in Figure 4.2. The toolbar contains convenient shortcuts for a number of common functions, such as saving and running the model. The functions of each button on the toolbar are

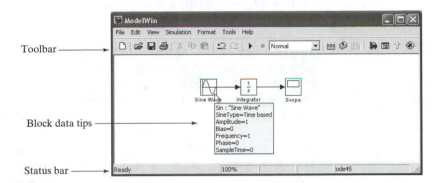

Toolbar

Block data tips

Status bar

FIGURE 4.2: Model window

TABLE 4.1: Model Window Toolbar Buttons

Button Icon	Function
	The **New model** button creates a new empty model window.
	The **Open model** button opens the **File:Open model** dialog box.
	The **Save model** button saves the model. If the model has not been previously saved, the **Save model** button opens the **File:Save as** dialog box.
	The **Print** button opens the **File:Print model** dialog box.
	The **Cut** button deletes a selection and places it on the clipboard. The **Cut** button is dimmed when nothing is selected.
	The **Copy** button places the current selection on the clipboard. The **Copy** button is dimmed when nothing is selected.
	The **Paste** button places the contents of the clipboard in the model window. The **Paste** button is dimmed if nothing is on the clipboard.
	The **Undo** button reverses the effect of the last copy, cut, delete, or add editing operation. Simulink provides 101 levels of Undo operations.
	Redo reverses the last Undo operation.
	The **Start** button is a shortcut for model window menu choice **Simulation:Start**.
	The **Stop** button is a shortcut for model window menu choice **Simulation:Stop**.
	The **Build all** button is displayed if Real-Time Workshop is installed. See Chapter 15.
	The **Update diagram** button is displayed if Real-Time Workshop is installed. See Chapter 15.
	The **Build subsystem** button is displayed if Real-Time Workshop is installed. See Chapter 15.

TABLE 4.1: Model Window Toolbar Buttons (Cont)

Button Icon	Function
	The **Simulink library browser** button opens the Simulink library browser window. It is equivalent to selecting the Simulink library browser button on the MATLAB toolbar.
	The **Toggle browser** button opens a model browser pane in the model window. The browser provides a convenient means for navigating through complicated models, particularly those involving subsystems (Chapter 7).
	The **Go to parent** button causes the parent system to the current subsystem to be displayed as the active model window, another convenient means with which to navigate in complex models. This is not active in the main model window.
	The **Debug** button enables debugging using the graphical debugger user interface.

shown in Table 4.1. The status bar fields display the current model window zoom factor, the current simulation time (visible while the simulation is in progress), and the currently selected solver. The toolbar and status bar can be hidden using **View:Toolbar** and **View:Status** Bar from the model window menu bar.

4.3.1 Zooming

The model window menu bar **View** menu supplies commands to control the zoom factor of the model window. **View:Zoom In** allows you to view a portion of the model window in more detail, and **View:Zoom Out** presents a wider perspective. **View:Fit System to View** zooms the window so that the model fills the model window. If any block or signal line is selected, there will be a menu choice **View:Fit Selection to View** that zooms the window so that the selection fills the model window. Finally, **View:Normal (100%)** sets the zoom factor to 100%.

4.3.2 Block Data Tips

Block data tips (when enabled) are displayed in a pop-up window that appears when the cursor is stationary on a block for more than approximately one second. The Block data tips window can be configured to display a number of items of information about the block, such as block parameters and a user description character string. You can configure the contents of the Block data tips window using **View:Block Data Tips** submenu commands.

4.3.3 Simulink Block Library

The default method to access Simulink blocks is via the Simulink Library Browser (Figure 3.4). An alternative is to open the Simulink Block Library by right-clicking the Simulink Library icon in the Simulink Library Browser window and clicking the single menu choice **Open the 'Simulink' Library**. The Simulink Block Library is

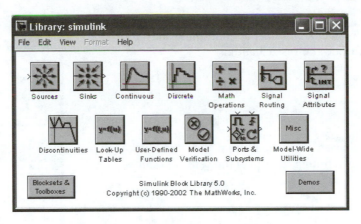

FIGURE 4.3: Simulink block library

a special type of model window called a *block library*, to be discussed in detail in Chapter 7. The Simulink Block Library is shown in Figure 4.3.

4.4   MANIPULATING BLOCKS

In Chapter 3, you learned how to drag a block from a block library to a model window and to flip a block. In addition to those basic operations, you can resize, rotate, copy, and rename blocks. In this section, we will discuss these and several other block manipulation operations. To prepare to practice the operations illustrated here, start Simulink, or, if Simulink is already started, open a new model window.

4.4.1   Resizing a Block

Frequently, resizing a block can improve the appearance of a model. To resize a block, proceed as follows:

Open a block library, and drag a block to the model window. Here, we're using a Gain block from the Math block library. We've set block parameter **Gain** to the value 1227.86. Because this value will not fit on the block icon, the displayed value is -K-.

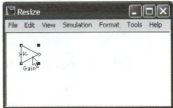

To resize a block, first select the block, causing the handles to appear.

Click the desired handle, and, continuing to depress the mouse button,

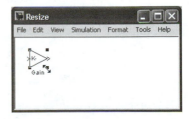

drag the handle to resize the block. Notice that the cursor changes shape, confirming that you have grasped the resize handle.

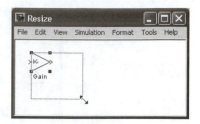

Release the mouse button. Now, the block icon is large enough to display the value of **Gain**.

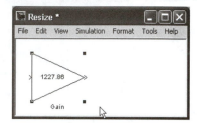

4.4.2   Rotating a Block

Occasionally, you will want to rotate a block. Select the block, then choose **Format:Rotate**.

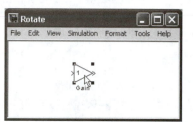

The block will rotate 90 degrees clockwise.

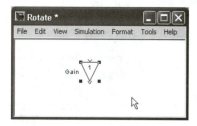

4.4.3   Copying a Block within a Model

You will frequently want to copy a block from within a model. For example, after you have resized a block, you will probably want all similar blocks to appear identical. Rather than attempting to resize each block, you can copy the block you resized.

To copy a block within a model, depress and hold the Control key, then click the block.

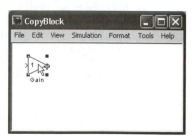

Drag the copy to the desired position.

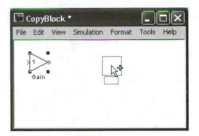

Release the mouse button, completing the copy operation.

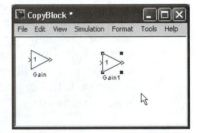

If you have a two- (or three-) button mouse, dragging by using the right mouse button is equivalent to dragging with the Control key depressed.

An alternative is to click the block, select **Edit:Copy** or the **Copy** button from the model window toolbar, or press Control-C. Then, choose **Edit:Paste** or the **Paste** button from the model window toolbar, or press Control-V.

4.4.4 Deleting Blocks

To delete a block, select the block, and then press the Delete key. An alternative is to select the block, then choose **Edit:Clear** from the model window menu bar or **Cut** from the model window toolbar. To delete the block and place it on the Clipboard, select the block, then choose **Edit:Cut** or the **Cut** button from the model window toolbar, or press Control-X.

4.4.5 Selecting Multiple Blocks

You can select multiple blocks and move, copy, or delete them as a group. There are two ways to select multiple blocks. The first is to depress and hold the Shift key while clicking each block in the group. The other method is to use a bounding box as described next.

Click and hold the mouse button outside the blocks.

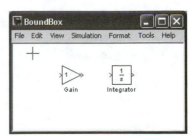

Drag the bounding box that appears so as to enclose all the desired blocks.

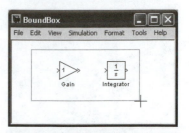

Release the mouse button. All blocks in the bounding box are now selected. Once the blocks are selected, the procedures for moving, copying, or deleting the entire group are the same as the corresponding procedures for a single block. For example, to delete the group, press the Delete key or choose **Edit:Cut** or **Edit:Clear** from the model window menu bar.

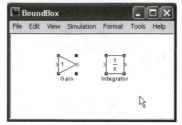

4.4.6 Changing a Block Label

Simulink supplies a default label for each block as you place the blocks in the model window. For example, the first Gain block will have the default name "Gain", the second gain block will be labeled "Gain1", and so on. Change the block label as described next.

Click the block label. An editing cursor will appear. You can position the cursor anywhere in the label by clicking, and you can move the cursor by using the cursor movement keys.

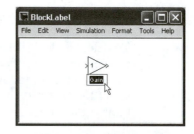

To replace a label, select it, then double-click it. The label will be highlighted as shown here.

Type the new label.

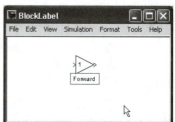

To create a multiple-line label, press Return at the end of each line.

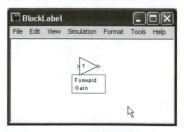

Click away from the block to accept the label.

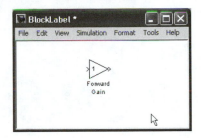

Each block in a window must have a unique name of at least one character.

4.4.7   Changing Label Location

You can move the block label from below to above the block as described below.
Select the block.

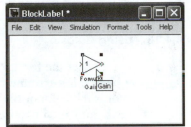

Choose **Format:Flip Name**. An alternative is to
click the name and drag it to the desired position.
If the block is rotated 90 degrees, the block label
will be to the left or the right of the block.

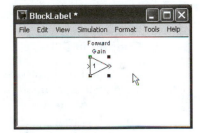

4.4.8   Hiding a Label

It is sometimes desirable to hide the name of a block. For example, because the
shape of a Gain block uniquely identifies the purpose of the block, and the value
of the gain is displayed on the block, a cluttered model might be improved by
hiding the names of the gain blocks. To hide the name, select the block, and choose
Format:Hide Name. Note that this does not change the name of the block in any
way—it simply makes the name invisible. If the name of a block is hidden, when
the block is selected, **Format:Hide Name** is replaced by **Format:Show Name** on the
Format pull-down menu.

4.4.9   Adding a Drop Shadow

If you want to call special attention to a block, you can apply a drop shadow.

Select the block.

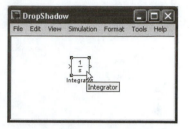

Choose **Format:Show Drop Shadow** from the model window menu bar. If a block is configured with a drop shadow, the menu choice will change to **Format:Hide Drop Shadow** for that block, permitting you to remove the drop shadow if desired.

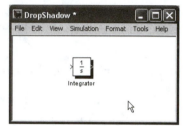

4.4.10   Using Color

You can set the foreground and background colors for each block or a selected set of blocks. Foreground color is the color of a block's outline and labels. Background color is the color of the body of the block. To change the color of a block, select the block, and then choose **Format:Foreground Color** or **Format:Background Color**. You can also change the color of the model window background using **Format:Screen Color**.

4.4.11   Configuring Blocks

In Chapter 3, we discussed setting block parameters. In the example in Chapter 3, the numeric parameters were set to constants. However, configuration parameters do not have to be constants. They can be any valid MATLAB expression and may use variables that will be defined in the MATLAB workspace when the model is executed. This is a very useful capability. Later, we'll discuss executing a Simulink model from a MATLAB script. By making certain parameters variables, the parameters can be changed by the script.

In addition to setting block parameters using the block parameters dialog box, you can set additional block characteristics using **Edit:Block Properties** (also available by right-clicking a block). The Block Properties dialog box consists of three tabbed pages: **General**, **Block Annotation**, and **Callbacks**. The Block Properties dialog box **Block Annotation** page for a Sine Wave block is shown in Figure 4.4 along with the corresponding block. The three tabbed pages contain five fields, as described in Table 4.2.

4.5   SIGNAL LINES

In Chapter 3 we discussed drawing signal lines using the Simulink automatic routing capability and drawing signal lines in segments. In this section we'll discuss techniques to edit signal lines.

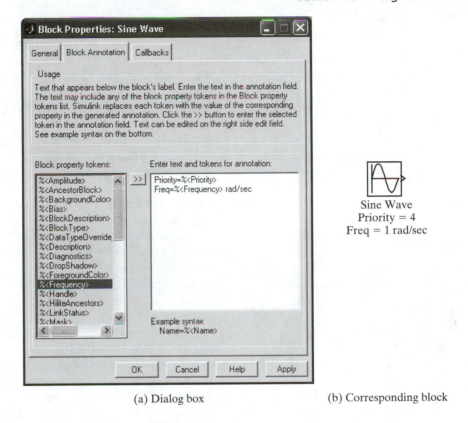

(a) Dialog box   (b) Corresponding block

FIGURE 4.4: Block Properties dialog box **Annotation** tab

4.5.1 Moving a Segment

To move a line segment, click the segment.

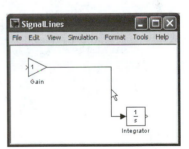

The cursor will change shape.

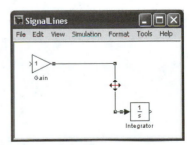

TABLE 4.2: Block Properties Dialog Box Fields

Page	Field	Purpose
General	Description	A user-specified character string containing any desired descriptive information. This field may be displayed in the Block Data Tips dialog box.
	Priority	A user-selectable value indicating the priority of the block in the simulation. You can set the priority of evaluation of all blocks in a model. There is no guarantee as to the order of evaluation for blocks with no priority assigned and no guarantee as to the order of evaluation for blocks of the same priority. Blocks with lower values of priority are evaluated before blocks with higher values of priority.
	Tag	A user-selectable character string associated with a block.
Annotation		A set of tokens from the **Block property tokens** list and text. Use the newline command (\n) to force a carriage return.
Callbacks		This page provides a convenient means with which to assign callback functions to a block. The list of available callbacks is in the **Callback functions list**. Details on callback functions are presented in Chapter 9.

Keeping the mouse button depressed, drag the segment to the desired location.

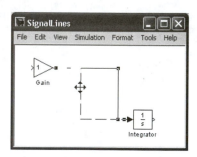

Release the mouse button.

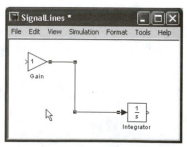

4.5.2   Moving a Vertex

To move a vertex, start by clicking the vertex.

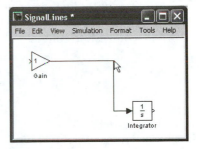

The cursor will change shape to a circle.

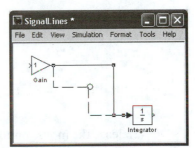

Drag the vertex to the desired location.

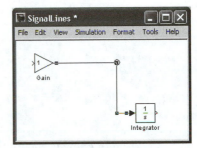

Release the mouse button to complete the operation.

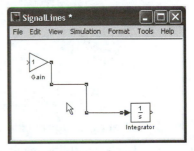

4.5.3   Deleting a Signal Line

To delete a signal line, select the signal line by clicking it. Press the Delete key, or choose **Edit:Clear** or **Edit:Cut** from the model window menu bar.

4.5.4 Splitting a Signal Line

To split a line, first select the line. Depress the Shift key, then click the line at the point at which you wish to split it.

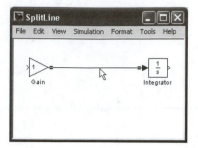

The cursor will change shape to a circle, and the segment will split into two segments.

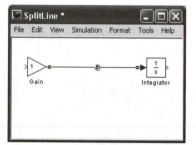

Drag the new vertex to the desired location.

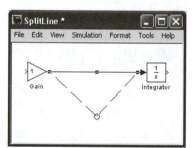

Release the mouse button to complete the operation.

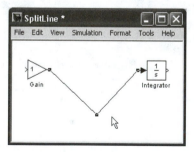

4.5.5 Labeling a Signal Line

Each signal line may have a label. The label can be positioned at either end of the signal line and on either side. Signal line labels, unlike block labels, do not have to be unique.

To add a label to a signal line, double-click the line, causing an editing cursor to appear near the line. Be sure that you click the line itself, not just near the line. Double-clicking away from a line will result in an annotation (to be discussed later) rather than a signal line label.

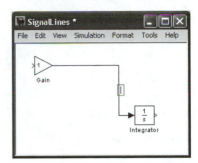

Enter the label. As with block labels, press Return at the end of each line to enter a label consisting of more than one line of text.

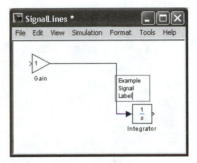

Click away from the line to complete entering the label. The label will snap to a position near the center of the line.

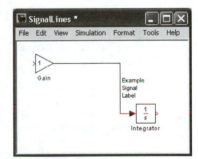

4.5.6 Moving or Copying a Signal Line Label

You can move a signal line label to either end or the middle of the signal line. To move a signal line label, click the label.

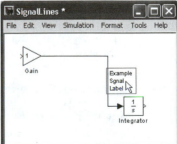

Drag the label to the desired location near the signal line.

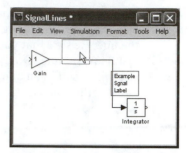

Release the mouse button. The label will snap into position.

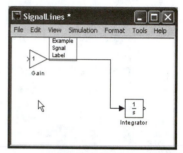

To copy a signal line label, depress the Control key while dragging the label to the new position.

If your mouse has two or three buttons, dragging with the right mouse button is an alternative to dragging while depressing the Control key.

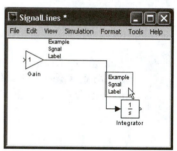

4.5.7   Editing a Signal Line Label

You can edit a signal line label using the same techniques used to edit a block label. All occurrences of the label will be changed.

If there are multiple occurrences of a signal line label, you can delete a single occurrence. Depress the Shift key, then click on the occurrence you wish to delete. Press the Delete key to complete the process.

To delete all occurrences of a signal line label, delete all characters in one instance of the label. When you click away from the label, all occurrences will be removed.

4.5.8   Signal Label Propagation

Signal line labels can propagate through several blocks in the Connections block library. Among these are the Mux and Demux, Goto and From, and Inport and Outport blocks. Signal line propagation provides an accurate means with which to determine the exact content of a signal line. This can make a model easier to understand and can also be useful in debugging. The process is illustrated next.

Here, we have a model in which two scalar signals produced by Constant blocks are combined to form a vector signal by a Mux block. The Demux block splits the vector signal into two scalar signals that are displayed by Display blocks. Notice that the model is shown before running the simulation (the Display blocks initially display 0). First, label the source signal lines. Here, the outputs of the constant blocks KA and KB are labeled A and B.

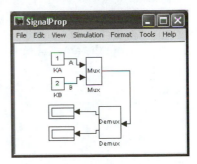

Label each signal line onto which you want the labels propagated with the single character <. Here, we configured all three signal lines to propagate the labels. Note that the > character is automatically added by Simulink.

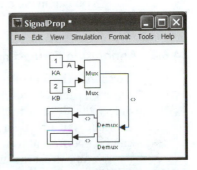

Choose **Edit:Update Diagram** from the menu bar of the model window. Here, the signal line leaving the Mux block contains two signals, so the label is changed to <A,B>, showing both. The Demux block separates the vector signal into components, so the signal lines leaving it each contain only one signal, as shown by the labels. In both cases, the propagated labels arc enclosed in angle brackets to distinguish them from simple labels.

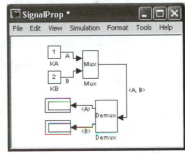

4.6   ANNOTATIONS

You can add annotations to a model to make it easier to understand. You can also change the font used in an annotation to add emphasis.

4.6.1   Adding Annotations

Double-click at the location where you want the center of an annotation to be. An editing cursor will appear.

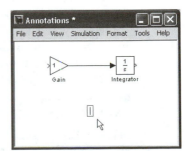

Enter the annotation. Press Return at the end of each line of a multiple-line annotation. Click away from the annotation to complete the process.

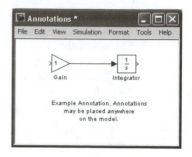

You can move and copy annotations using the same procedures used to move and copy blocks.

4.6.2   Annotation Formatting

To change the font of an annotation, select the annotation. Choose **Format:Font** from the menu bar of the model window. A font selection dialog box will be displayed. Select the desired font, then press **OK**; then click away from the annotation. All characters in a particular annotation will be the same font, but different annotations can be in different fonts.

You can set the justification for an annotation using **Format:Text alignment** to **Left**, **Centered**, or **Right**.

4.7   ADDING SOURCES

The inputs to a model are called *sources*, located in the Sources block library. A source block has no inputs and at least one output. Detailed documentation for each block in the Sources block library is available via the online Help system. In this section, we will mention several of the more commonly used blocks in the Sources block library. Then, we will briefly discuss From Workspace and From File blocks that allow you to create any time-dependent signal you can describe mathematically in MATLAB, and then use that signal as a Simulink input.

4.7.1   Common Sources

Many of the input signals used in modeling dynamical systems are available in the Sources block library. The Constant block produces a fixed constant signal, the magnitude of which is set in the block dialog box and displayed on the block icon. The Step block produces a step function. You can set the time of occurrence of the step and the signal magnitude before and after the step. There is a Sine Wave block for which you can set the amplitude, phase, and frequency. The Signal Generator block can be set to produce sine, square, or sawtooth. More complex signals can be generated by combining the signals from multiple source blocks using a Sum block.

EXAMPLE 4.1   Unit impulse function

A signal that is useful in determining the behavior of dynamical systems is the unit impulse, also known as the delta function or Dirac delta function. (For a detailed discussion, see

Meirovitch [1].) The unit impulse $\delta(t - a)$ is defined to be a signal of zero duration, having the properties:

$$\delta(t - a) = 0, \qquad t \neq a$$

$$\int_{-\infty}^{\infty} \delta(t)dt = 1$$

Although the unit impulse is a theoretical signal that cannot exist, it is a close approximation to real impulse signals that are common. Physical examples are collisions, such as a wheel hitting a curb or a bat hitting a ball, or near-instantaneous velocity changes, such as firing a bullet from a rifle. Another use for the unit impulse is the assessment of a system's dynamics. The motion of a system forced by a unit impulse is due purely to the dynamics inherent in the system. Thus, you can use the impulse response of a complex system to determine its natural frequencies and modes of vibration.

You can approximate a unit impulse function using two Step blocks and a Sum block, as shown in Figure 4.5. The idea is to produce, at the desired time a, a very

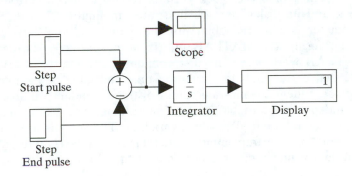

FIGURE 4.5: Generating a unit impulse function

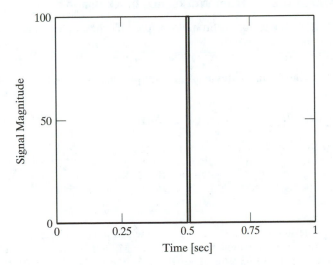

FIGURE 4.6: Unit-impulse signal

short duration d pulse of a magnitude M, such that $Md = 1$. The trick is in deciding on the proper value of d. It must be short relative to the fastest dynamics of the system. If it is too short, it can cause numerical problems, such as excessive round-off error. If it is too long, it will not adequately simulate a true impulse. Usually, an acceptable compromise can be found with a little experimentation.

The model in Figure 4.5 is set to simulate a unit impulse at 0.5 seconds with a pulse of 0.01 seconds duration and magnitude of 100. The Step block labeled Step Start pulse is configured as follows: **Step time** is 0.5, **Initial value** is 0, and **Final value** is 100. The Step block Step End pulse is configured as follows: **Step time** is 0.51, **Initial value** is 0, and **Final value** is 100. The simulation is configured to stop at 1 second. A plot of the output of the Sum block (sent by the Scope block to the MATLAB workspace) is shown in Figure 4.6. The Integrator block computes the time integral of the output of the Sum block, which is displayed using a Display block, and which has the desired value (1).

■

4.7.2 From Workspace Block

The From Workspace block permits you to design a custom input signal. The block and its dialog box are illustrated in Figure 4.7. The block configuration parameter is a matrix table with the default value [T,U]. The input must be in the form of a MATLAB matrix, using variables currently defined in the MATLAB workspace. The first column of the matrix is the independent variable that corresponds to simulation time and must be monotonically increasing. The subsequent columns are values of the dependent variables corresponding to the independent variable in the first column. The block will produce as many outputs as there are dependent variables. The outputs are produced by linearly interpolating or extrapolating in the table. Check boxes **Interpolate data** and **Hold final data value** modify the behavior of the block. For details, refer to the block help screen.

■

EXAMPLE 4.2   From Workspace block usage

To illustrate the use of the From Workspace block, suppose that we wish to generate a signal defined as

$$u(t) = t^2$$

Listing 4.1 illustrates an M-file that produces a suitable input table.

Listing 4.1: M-file to create input table for From Workspace block

```
%Generate a signal for a From Workspace block
t = 0:0.1:100 ;    %Independent variable
u = t.^2      ;    %Dependent variable
A = [t',u']   ;    %Form table
```

Note that the M-file must be saved with a name different from the Simulink model file. To use this table, you must first execute the M-file from within MATLAB, creating the table A. Configure the From Workspace block by replacing [T,U] in the From Workspace dialog box with the name of the table (A).

From
Workspace

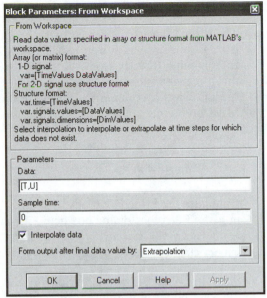

(a) From Workspace Block   (b) From Workspace dialog box

FIGURE 4.7: From Workspace block and dialog box

Set Matrix table to A, the name of the table created in the MATLAB workspace. Click Close.

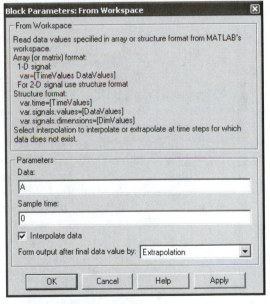

The block icon will display the name of the table.

From
Workspace

4.7.3   From File Input Block

The From File block is similar to the From Workspace block. The primary difference is that the matrix is stored in a file in MATLAB matrix file (.mat) format rather than coming directly from the MATLAB workspace. An additional difference is that the signals are stored in rows rather than in columns. You can produce a file (examp.mat) containing the matrix produced by the M-file in Listing 4.1 using the MATLAB commands:

```
B = A'
save examp B
```

Configure the From File block by setting the full filename (for example, examp.mat) in From File block dialog box field **File name**.

4.8   ADDING SINKS

Sinks provide the means to view or store model data. The Scope and XY Graph blocks produce plots of model data, and the Display block produces a digital display of the value of its input. The To Workspace block saves a signal to the MATLAB workspace, and the To File block saves a signal in MATLAB .mat file format. The Stop block causes a simulation to stop when its input is nonzero. A detailed reference for each of these blocks is available in the online Help system. We will discuss the Scope block and XY Graph block in some detail in this section.

4.8.1   Scope Block

The Scope block emulates an oscilloscope. The block shows a segment of the input signal, which may be scalar or vector. Both the vertical range (y-axis) and horizontal range (Time, on the x-axis) can be set to any desired values. The vertical axis displays the actual value of the input signal. The horizontal axis scale always starts at zero and ends at the value specified as time range. So, for example, if the horizontal range is 10 and the current time is 100, the input data for the period 90 to 100 is displayed, although the x-axis labels will still be 0 to 10. The Scope block is intended primarily for use during a simulation, but the block has the capability to produce a printed copy of the image. Additionally, the Scope block will send the signals it plots to the MATLAB workspace for further analysis or plotting using, for example, the MATLAB plot command or the MATLAB simplot command.

You can place a Scope block in a model without connecting a signal line to the input of the Scope block, and configure the block as a floating Scope block. A floating Scope block will use as its input any signal line that you click during the execution of a simulation.

Figure 4.8 illustrates a Scope block. Note that there is a toolbar that contains eleven icons along the top of the Scope block window. These buttons allow you to zoom in on a portion of the display, autoscale the display, save a configuration for future use, and open a scope properties dialog box. Described in Table 4.3 are the functions of each button.

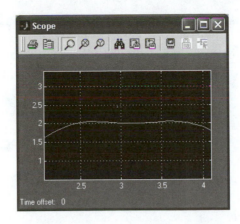

FIGURE 4.8: Scope block display before running simulation

TABLE 4.3: Scope Toolbar Buttons

Button Icon	Function
	Click the **Printer** button to print the scope plots.
	Click the **Parameters** button to open the Scope parameters dialog box. This dialog box allows you to set the default scales for the Scope block and to send the Scope data to the MATLAB workspace.
	The **Zoom** button allows you to enlarge a region of the display.
	The **Zoom X** button allows you to zoom in on a portion of the display without changing the vertical scale.
	The **Zoom Y** button allows you to zoom in on a portion of the display without changing the horizontal scale.
	Autoscale changes the vertical scale such that the lower limit is the same as the minimum value in the currently displayed signal, and the upper limit is the same as the maximum value in the currently displayed signal. You can click **Autoscale** during a simulation to rescale the display.
	Save axis makes the current scale the default for this Scope block. If you change the scale and then rerun the simulation without pressing Save axis first, the scale will revert to the current default when the simulation starts.
	Restore axis causes the Scope to return to the currently saved axis settings after zooming.
	The **Floating scope** button converts the Scope block into a Floating Scope block. Pressing the **Floating scope** button again returns the block to a normal Scope block.
	The **Lock Axes selection** button toggles axis selection for Floating Scope blocks. The button is not active for normal Scope blocks.
	The **Signal selection** button opens the signal selection dialog box for Floating Scope blocks. The button is not active for normal Scope blocks.

Zooming the Scope Display. Consider the model shown in Figure 4.9. The top Sine Wave block is configured to produce the signal $\sin t$, and the other Sine Wave block is configured to produce $0.4 \sin t$.

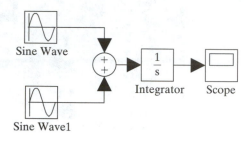

FIGURE 4.9: First-order system with sinusoidal input

Open the Scope block by double-clicking it. Run the simulation, resulting in the display shown here. At this point, you could rescale the display by clicking the **Autoscale** button. Instead, zoom in on the peak between 2 and 4 seconds, as follows.

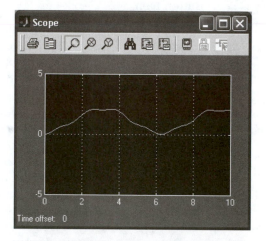

Click the **Zoom** button. Then, enclose the area you wish to zoom in on using a bounding box.

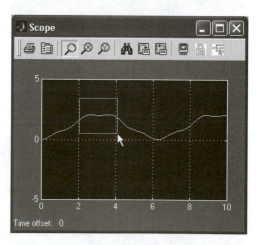

The Scope scales will change to include only the area you zoomed in on.

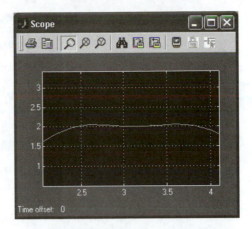

Scope Parameters Dialog Box.   The **Parameters** button opens the Scope Parameters dialog box, which has two pages. The **General** page (Figure 4.10) has fields to set the number of axes and time range and to control the spacing between plotted points. The **Data history** page (Figure 4.11) has fields to control the size of the scope data buffer and to send the displayed data to the MATLAB workspace.

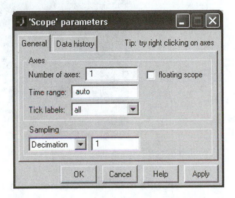

FIGURE 4.10: Scope parameters **General** page

The **General** page (Figure 4.10) has two sections. The Axes section controls the number and configuration of axes. **Number of axes** controls the number of axes displayed in the scope window and the number of inputs to the Scope block. There will be one input for each axis. For example, setting **Number of axes** to 2 produces the scope block and display shown in Figure 4.12. Checking **floating scope** converts the Scope block to a Floating scope block, which has no inputs. A floating scope block displays a signal line that is selected during the execution of a simulation. Note that in order to use a Floating scope block with many Simulink blocks, it is necessary to disable **Signal storage reuse** in the **Simulation:Parameters** dialog box Advanced page, to be discussed in Section 4.10. **Time range** controls the scale of the time axis. If set to auto, the scale will range from zero to the final simulation time. If set to a number greater than zero, the time range will be zero to the value entered.

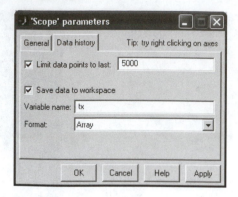

FIGURE 4.11: Scope parameters **Data history** page

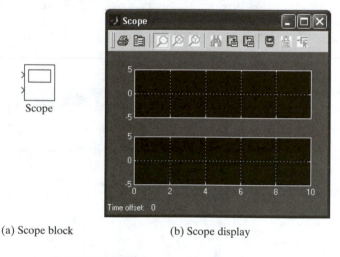

(a) Scope block   (b) Scope display

FIGURE 4.12: Scope block with two axes

Tick labels is a drop-down menu that can be set to all to place time ticks on each time axis, none, or bottom axis only for ticks on the x-axis only.

The Sampling section of the **General** page contains a drop-down list containing two choices: **Decimation** and **Sample time**. If **Decimation** is selected, the corresponding data field is set to a decimation factor that must be an integer. If **Decimation** is selected and set to 1 (the default), every point in the block input is plotted. If **Decimation** is set to 2, every other point is plotted, and so on. If **Sample time** is selected, the absolute spacing between plotted points must be entered in the data field.

The Scope block stores the input points in a buffer. The size of the buffer can be set using the **Data history** page. Check **Limit points to last** and enter a value to specify the size of the buffer (default is 5000). Autoscaling, zooming, and saving scope data to the workspace all work with this buffer. Thus, if **Limit rows to last** is set to 1000, and the simulation produces a total of 2000 points, only the final 1000 points are available when the simulation stops.

Although you can print a scope display, you have little control over its appearance. Additionally, the Scope printing capability is not intended to support placing the plot in a word-processing document. However, you can send the Scope data to the MATLAB workspace, and use the extensive plotting capabilities MATLAB provides. To send the Scope data to the MATLAB workspace, select **Save data to workspace** and enter the name of a MATLAB variable. When the simulation stops, the data displayed on the Scope will be stored in the MATLAB variable. Drop-down menu **Format** controls the type and contents of the MATLAB variable. If **Matrix** is selected, there will be one column for the time values and one column for each signal input to the Scope block. Thus, if the signal entering the Scope block is a vector signal with two components, the MATLAB variable will be a matrix with three columns and a number of rows equal to the number of time points displayed on the Scope. If **Structure** is selected, the output will be a MATLAB structure, but no time vector will be stored. Choose **Structure with time** to add a time vector to the structure. If you send the Scope data to the MATLAB workspace, you may use the MATLAB `simplot` command to produce an editable MATLAB plot that resembles the Scope display.

Setting *y*-Axis Limits.   To set *y*-axis limits, right-click the scope display. In the dialog box that appears, choose **Axis properties**. Another dialog box will appear. This dialog box has fields in which to enter the axis limits and to specify an axis label. The default special string %<`SignalLabel`> will display the signal label of the input signal, if the signal has a label. The label appears centered at the top of the axis.

4.8.2   XY Graph

The XY Graph block produces a graph identical to a graph produced by the MATLAB command `plot`. The XY Graph accepts two scalar inputs. You must configure the horizontal and vertical ranges using the block dialog box. The XY Graph block dialog box is illustrated in Figure 4.13. The top input port is the *x*-input, and the bottom input port is the *y*-input.

4.9   DATA TYPES

Most programming languages, including MATLAB, provide multiple data types. A data type is a computer representation of a number, both the amount of storage allocated to the number and the format of the representation. MATLAB and Simulink provide long and short integers, single-precision and double-precision floating point, and Boolean data types. In Table 4.4, the available data types are listed. The size (in bytes) of each data type is dependent upon the computer and operating system used. See the MATLAB online documentation for details.

The default data type for all Simulink signals and block parameters is double precision floating point. Therefore, it is never necessary for you to use the other available data types. Often, however, using the other data types can improve the fidelity of a simulation. For example, a model of a control system that uses Boolean data, analog to digital converters, or finite word length

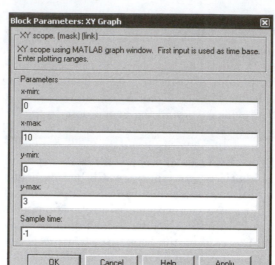

FIGURE 4.13: XY Graph block dialog box

TABLE 4.4: Simulink Data Types

Type	Meaning
double	Double-precision floating point: This is the default Simulink data type.
single	Single-precision floating point: Usually requires one-half the storage of a double-precision floating point.
int8	Signed 8 bit integer.
uint8	Unsigned 8 bit integer.
int16	Signed 16 bit integer.
uint16	Unsigned 16 bit integer.
int32	Signed 32 bit integer.
uint32	Unsigned 32 bit integer.
boolean	Logical data: False is 0, True is 1. If the Data Type Conversion block is set to **boolean**, it converts any nonzero input to 1.

filters can be improved through the use of appropriate data types. Also, if a model will be used with Real-Time Workshop, significant performance improvements can result from using more compact data types, where possible. Additionally, the Combinatorial Logic and Logical Operator blocks (see Chapter 6) can be configured via the **Simulation:Parameters** dialog box to require Boolean signals.

In a Simulink model, signals and block parameters can be set to any data type. To set a block parameter to a particular data type, use the type-casting notation type(value), where type is any of the data types listed in Table 4.4, and value is the value assigned. For example, to set a block parameter to a 16-bit integer with the value 25, enter in the block dialog box field for the parameter.

```
int16(25)
```

To convert a signal to a particular data type, use the Data Type Conversion block from the Signal Attributes block library.

You can configure a model such that the data types at each block port are displayed. Choose **Format:Port Data Types** from the model window menu bar.

4.10   CONFIGURING THE SIMULATION

A Simulink model is essentially a computer program that defines a set of differential and difference equations. When you choose **Simulation:Start** from the model window menu bar, Simulink solves that set of differential and difference equations numerically, using one of its differential equation solvers. Before you run a simulation, you may set various simulation parameters, such as the starting and ending time, simulation step size, and various tolerances. You can choose among several high-quality integration algorithms. You can also configure Simulink to acquire certain data from the MATLAB workspace, and to send simulation results to the MATLAB workspace.

To illustrate the basic idea, consider the model shown in Figure 4.14. This model represents the differential equation

$$\dot{x} = 2$$

We set the Integrator block **Initial condition** field to 1. Next, we choose **Simulation:Parameters** from the model window menu bar, and set **Start time** to 0, and set **Stop time** to 5. Then choose **Simulation:Start** from the model window menu bar. Simulink will numerically evaluate the value of the integral to solve

$$x(\tau) = 1 + \int_0^\tau 2\,dt$$

and plot the value of $x(\tau)$ along the interval from 0 to 5. A numerical integration algorithm that solves this kind of problem (frequently called an initial value problem) is referred to as an *ordinary differential equation solver*, or just *solver* for the sake of brevity.

To set simulation parameters, choose **Simulation:Parameters** from the menu bar of the model window, which opens the Simulation parameters dialog box

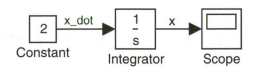

FIGURE 4.14: Simulink model of first-order system

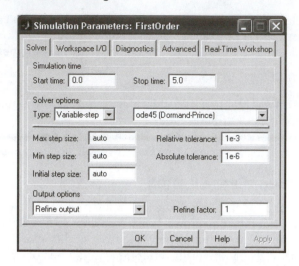

FIGURE 4.15: Simulation parameters dialog box

(Figure 4.15). The Simulation parameters dialog box contains four tabbed pages: **Solver**, **Workspace I/O**, **Diagnostics**, and **Advanced**. If Real-Time Workshop is installed, there will be a fifth tabbed page titled **Real time Workshop**. The **Solver page**, illustrated in Figure 4.15, selects and configures the differential equation solver. The **Workspace I/O** page contains optional parameters that permit you to acquire simulation initialization data from the MATLAB workspace and to send certain simulation data to the MATLAB workspace. The **Diagnostics** page is used to select diagnostic modes, which are useful for troubleshooting certain simulation problems. We'll discuss each of these four pages in detail, next.

4.10.1 Solver Page

The **Solver** page consists of three sections. The first, Simulation time, contains fields in which to enter the start and stop times. **Start time** defaults to 0, and **Stop time** defaults to 10. The Solver options section contains fields to select the differential equation solver (numerical integration algorithm) and to set parameters that control the integration step size. The solvers are grouped in two categories: variable-step and fixed-step. Several different integration algorithms are available for each category. If a variable-step solver is chosen, there are fields to select the maximum integration step size, the initial integration step size, and absolute and relative tolerances. If a fixed-step solver is chosen, there is a single field in which to enter the step size. The Output options section controls the time spacing of points in the simulation output trajectory.

Solver Type. Simulink provides several ordinary differential equation solvers. The majority of these solvers are the result of recent numerical integration research, and they are among the fastest and most accurate methods available. Detailed descriptions of the algorithms are available in the paper by Shampine [2] and Hosea [3].

It is generally best to use the variable-step solvers, as they continuously adjust the integration step size to maximize efficiency, while maintaining a specified accuracy. The Simulink variable-step solvers can completely decouple the integration step size and the interval between output points, so it is not necessary to limit the step size to get a smooth plot or to produce an output trajectory with a predetermined fixed step size. The available solvers are listed in Table 4.5. We will discuss them in more detail in Chapter 13.

TABLE 4.5: Simulink Solvers

Solver	Characteristics
ODE45	Excellent general-purpose one-step solver. Based on the Dormand–Prince fourth/fifth-order Runge–Kutta pair. ODE45 is the default solver, and it is usually a good first choice.
ODE23	Uses the Bogacki–Shampine second/third-order Runge–Kutta pair. It sometimes works better than ODE45 in the presence of mild stiffness. It generally requires a smaller step size than ODE45 to get the same accuracy.
ODE113	Variable-order Adams–Bashforth–Moulton solver. Because ODE113 uses the solutions at several previous time points to compute the solution at the current time point, it may produce the same accuracy as ODE45 or ODE23 with fewer derivative evaluations, and thus perform much faster. It is not suitable for systems with discontinuities.
ODE15S	Variable-order multistep solver for stiff systems. It is based on recent research using numerical difference formulas. If a simulation runs extremely slowly using ODE45, try ODE15S.
ODE23S	Fixed-order one-step solver for stiff systems. Because ODE23S is a one-step method, it is sometimes faster than ODE15S. If a system appears to be stiff, it is a good idea to try various stiff solvers to determine which one performs best.
ODE23T	This is a one-step solver for stiff systems, based on trapezoidal integration. ODE23T and ODE23TB are variants of the same method. ODE23T is somewhat faster but also less stable.
ODE23TB	This is a variant of ODE23TB using backward differentiation formulas for error estimation. This version is more stable than ODE23T but is also somewhat slower.
Discrete	This is a special solver for systems that contain no continuous states.
ODE5	It is a fixed-step version of ODE45.
ODE4	Class fourth-order Runge–Kutta formulas using a fixed step size.
ODE3	Fixed-step version of ODE23.
ODE2	Fixed-step second-order Runge–Kutta method, also known as Heun's method.
ODE1	Euler's method using a fixed step size.

Output Options.   The Output options section of the **Solver** page works in conjunction with the variable-step solvers to control the spacing between points in the output trajectory. Output options do not apply to the fixed-step solvers. The **Output options** field contains a list box with three choices: `Refine output`, `Produce additional output`, and `Produce specified output only`.

Choose `Refine output` to force the solver to add intermediate points between the solution points for successive integration steps. Simulink computes the intermediate points using interpolation, which is much faster than using reduced integration step size. `Refine output` is a good choice if the output trajectory needs to appear smoother, but there is no need for a fixed spacing between points. If `Refine output` is chosen, there will be an additional input field labeled **Refine factor**. **Refine factor** must be an integer. Simulink divides each integration step into **Refine factor** output steps, so, for example, if **Refine factor** is set to 2, the midpoint of each integration step will be added to the output trajectory.

`Produce additional output` permits you to force Simulink to include certain time points in the output trajectory, in addition to the solution points at the end of each integration step. If `Produce additional output` is selected, there will be an additional field labeled **Output times**. This field must contain a vector listing the additional times for which output is requested. For example, if it is necessary to include the output at 10-second intervals, and the value of **Start time** is 0 and **Stop time** is 100, **Output times** should contain [0:10:100].

Choose `Produce specified output only` if it is necessary to produce an output trajectory containing only specified time points. For example, you may wish to compare several trajectories to evaluate the effect of changing a parameter. If `Produce specified output only` is selected, there will be an additional field labeled **Output times**, which must contain a vector of the desired output times.

4.10.2   Workspace I/O Page

The **Workspace I/O** page (Figure 4.16) permits you to acquire simulation input from the MATLAB workspace, and to send output directly to the MATLAB workspace. The page consists of three sections: Load from workspace, Save to workspace, and Save options. We'll discuss each section in detail.

Simulink Internal State Vectors.   Before discussing the Load from workspace and Save to workspace sections of the **Workspace I/O** page, we should briefly discuss Simulink internal state variables. A Simulink model can be thought of as a set of simultaneous first-order, possibly nonlinear, differential and difference equations. In addition to the state variables associated with each integrator block, there are implicitly specified state variables associated with transfer function blocks, state-space blocks, certain nonlinear blocks, certain discrete blocks, and many of the blocks in the Extras block library. It is frequently useful to have access to a model's state variables, and Simulink provides mechanisms to facilitate this. Use of the **Workspace I/O** page is probably the easiest method with which to access a model's state variables. Accessing a model's state variables, and, in particular, identifying all of a model's state variables, are discussed further in Chapter 8.

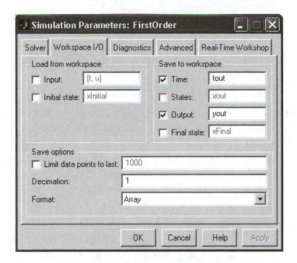

FIGURE 4.16: **Workspace I/O** page

Load From Workspace.   Selecting the **Input** check box causes Simulink to take the input time points and values of the input variables from the MATLAB workspace. **Input** works in conjunction with the Inport block found on the Ports and Subsystems block library. Inport blocks may be configured to accept scalar or vector data. Set the name of the time and input matrices in the Input field. The first matrix (default name t) is a column vector of time values, and the second matrix (default name u) consists of one column for each input variable, with a row corresponding to each row in the time matrix. If there is more than one Inport block, the columns of the input matrix are ordered corresponding to the number assigned to the Inport blocks. So, the first column corresponds to the lowest numbered Inport block, and the last column corresponds to the highest numbered Inport block. For a vector Inport block, there must be one column in the u matrix for each element of the input vector.

Selecting the **Initial state** check box of the Workspace I/O forces Simulink to load the initial values of all internal state variables from the MATLAB workspace. All Simulink state variables have default initial values, in most cases 0. States associated with Integrator blocks may be initialized to any value using the block's dialog box. Specifying initial states on the Workspace I/O page overrides any default initialization values, including initial values set in an Integrator block's dialog box. **Initial state** sets initial value of a model's state vector to the values in the specified input vector (default name xInitial), which must be defined in the MATLAB workspace when the simulation begins. The initialization vector must be of the same size as the model's state vector but may be either a row or column vector.

Save to Workspace.   The Save to workspace section contains four fields, each activated with a check box. **Time** sends the independent variable to the specified workspace matrix (default name tout). **States** sends all of the model's state variables to the specified MATLAB workspace matrix (default name xout). **Output** works in conjunction with Outport blocks in a manner analogous to Inport blocks, discussed above. **Final state** saves the final value of the model's state vector in the specified matrix (default name xFinal) in the MATLAB workspace. The output of **Final state** is a suitable initial vector for **Initial state** and may be used to restart a model at the final point of a previous simulation.

Save Options. The Save options section of the **Workspace I/O** page contains three fields, and it works in conjunction with the Save to workspace section. The first field, **Limit rows to last**, sends at most the specified number of points to the MATLAB workspace. So, for example, if **Limit rows to last** is checked and set to the default value of 1000, at most the final 1000 points will be sent to the workspace. **Decimation** sets the interval between points sent to the MATLAB workspace. If **Decimation** is set to 1, every point will be sent to the workspace. If **Decimation** is set to 2, every other point will be sent to the workspace, and so on. **Decimation** must be set to an integer value. The final field controls the format of the output variables in the MATLAB workspace. The default, Matrix, saves the data in a standard MATLAB matrix. There are also options for saving the data in the form of MATLAB structures with or without a time vector.

4.10.3 Diagnostics Page

The Diagnostics page (Figure 4.17) allows you to select the action taken for various exceptional conditions and also includes options to control automatic block output consistency checking, bounds checking, and simulation profiling. There are three choices for the response to each of the exception conditions. The first choice, None, instructs Simulink to ignore the corresponding exception. The second choice, Warning (the default choice), causes Simulink to issue a warning message each time the corresponding exception occurs. The final choice, Error, causes Simulink to abort the simulation and issue an error message whenever the corresponding

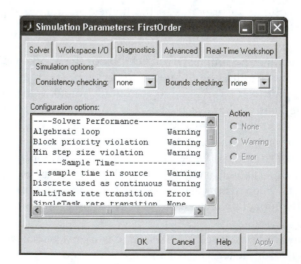

FIGURE 4.17: **Diagnostics** page

exception occurs. The exceptions are detected when a model is executed. A brief explanation for each exception is available from the Simulink Help browser. We'll briefly discuss a few of the more common exceptional conditions here.

An **Algebraic loop** is an exception in which a block's input at a given instant of time is dependent on the same block's output at the same instant of time. Algebraic loops are troublesome, because they can significantly reduce the speed of a simulation, and in some cases can cause the simulation to fail to execute. A detailed discussion of algebraic loops is presented in Chapter 13. It is usually best to set the algebraic loop response to Warning. If an algebraic loop is discovered, and if performance is acceptable, change the response to None.

A **Min step size violation** occurs when the solver attempts to use an integration step size smaller than the minimum. It is not possible to change the minimum step size for any of the variable step size solvers. If this exception occurs, you can change to a higher-order solver, which in general will use a larger integration step size. Your other choice is to increase the absolute and relative tolerances on the **Solver** page. **Min step size violation** should always be set to Warning or Error, because it indicates the simulation is not producing the expected accuracy.

Unconnected inputs, outputs, and signal lines are almost always errors. Therefore, the corresponding options should be set to Warning or Error. If it is necessary to leave an input or output unconnected, use Ground and Terminator blocks.

A **Data overflow** exception occurs when the magnitude of a signal or internal variable exceeds the limit for its data type. Data overflow is almost always a serious error, and therefore, **Data overflow** should be set to Warning or Error.

Consistency checking is a debugging feature that detects certain programming errors in custom blocks. Consistency checking is not needed with standard Simulink blocks, and it causes a simulation to run much slower. Ordinarily, **Consistency checking** should be set to None.

4.10.4 Advanced Page

The **Advanced page** (Figure 4.18) allows you to control certain simulation options that affect the speed of execution and memory usage. Included are options to disable zero crossing detection and to relax the rules for Boolean-type checking in certain blocks. These options are set to either On or Off.

A number of Simulink blocks exhibit discontinuous behavior. For example, the output of the Sign block, located in the Math Operations block library, is 1 if its input is positive, 0 if its input is zero, and −1 if its input is negative. Thus, the block exhibits a discontinuity at zero. If a variable step size solver is in use, Simulink will adjust the integration step size when the input to a Sign block is approaching zero, so that the switch occurs at the right time. This process is called *zero crossing detection*. You can determine whether a block invokes zero crossing detection by referring to the Characteristics table for the block in the online Help system.

Zero crossing detection improves the accuracy of a simulation, but it can cause a simulation to run slowly. Occasionally, a system will fluctuate rapidly about a discontinuity, a phenomenon called *chatter*. When this happens, the progress of the simulation can effectively stop, as the integration step size is reduced to a very small value. If your model runs slowly and includes one or more blocks with intrinsic

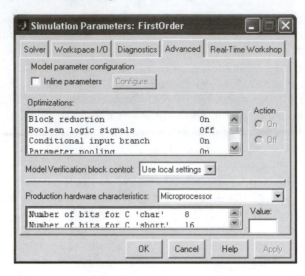

FIGURE 4.18: **Advanced** page

zero crossing detection, disabling zero crossing detection can significantly increase the speed of a simulation. However, this can also adversely affect the accuracy of a simulation, so it is best used only as a tool to verify that chatter is occurring. If disabling zero crossing detection dramatically improves the speed of a simulation, you should locate the cause of chatter and correct the problem.

As a consequence of the method by which Simulink optimizes storage, memory buffers are not maintained for every signal line. Ordinarily, this causes no problem. However, in order for floating Scope blocks to display a signal, there must be a buffer allocated for the signal line. Turn Signal storage reuse to Off to disable this optimization feature. This should only be done during development and debugging, as there is a performance and memory penalty.

A number of Simulink blocks, for example logic blocks, expect Boolean signals and will produce an error message if the input signals are not Boolean. Setting Boolean logic signals to Off suppresses these error messages. This feature is provided to ensure compatibility with earlier versions of Simulink, which had only one data type, double-precision floating point. If Boolean logic signals is set to Off, logic blocks will accept floating point signals. In this case, the value 0.0 is interpreted as False, and anything else is interpreted as True.

4.11 RUNNING A SIMULATION

You can control the execution of the model using the **Simulation** pull-down menu on the model window menu bar or by using the model window toolbar. To start the simulation, select **Simulation:Start** or click the **Start** button. You can stop the simulation at any time using **Simulation:Stop** or the **Stop** button on the model window toolbar. Choose **Simulation:Pause** to temporarily halt execution. Then, choose **Simulation:Continue** to resume or **Simulation:Stop** to permanently halt execution of the model. It is also possible to run a simulation from the MATLAB command line; this method will be discussed in Chapter 8.

While the simulation is executing, you can change many parameters. For example, you can change the gain of a Gain block, choose a different solver, or change integration parameters, such as minimum step size. You can also select a signal line that will become the input to a floating Scope block. This permits you to check various signals as the simulation progresses.

4.12 PRINTING A MODEL

There are a variety of options for printing Simulink models. The simplest is to send the model directly to a printer. However, it is also possible to embed the model in a word-processing document or other file, either by copying it to the clipboard (Microsoft Windows) or saving it as an Encapsulated PostScript file.

4.12.1 Printing to the Printer Using Menus

The fastest way to get a printed copy of a model is to send it directly to the printer. On Microsoft Windows, you can choose **File:Print** or click the **Print** button on the model window toolbar. The Print Model dialog box (Figure 4.19) will be displayed. In addition to standard print controls, the Print Model dialog box contains an Options section that controls which parts of a hierarchical Simulink model (see Chapter 7 for information on hierarchical models) are printed. There are also options to produce a log of printout pages and to add frames to each page.

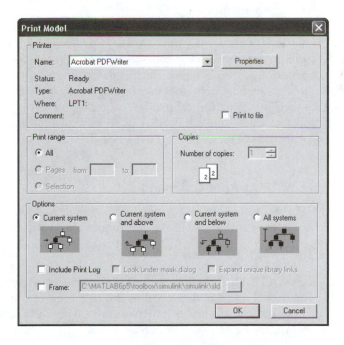

FIGURE 4.19: **Print Model** dialog box

Select **Frame** to add a background frame to each printout page. When **Frame** is selected, the frame file field is available. A frame file contains all information needed to produce the frame. Frame files can be created and edited using the frame file editor, which may be accessed by entering the command

```
frameedit
```

at the MATLAB prompt. For details on using the frame editor, enter the command

```
doc frameedit
```

at the MATLAB prompt.

4.12.2 Embedding the Model in a Document

Modern documentation applications such as word processors, presentation programs, desktop publishing programs, drawing programs, and even spreadsheets, allow you to insert Simulink model images in documents. Once you have embedded an image in a document, you can use the graphics capabilities of the documentation application to resize the image and to further annotate the model. Simulink model images may be embedded as bitmaps, as Windows metafiles, or as Encapsulated PostScript files. To embed an image as a metafile using Microsoft Windows, select **Edit:Copy Model** from the model window menu bar. The entire model window will be copied to the clipboard. Next, make the target document active, and use the procedure for the documentation application to embed the image. For most Microsoft Windows programs, the procedure is to choose **Edit:Paste Special**, then select Metafile (sometimes listed as Picture).

To embed the image of a Simulink model in a document in Encapsulated PostScript form, first save the image as an Encapsulated PostScript file using the print command, which will be discussed shortly. There appears to be no standard menu convention for embedding Encapsulated PostScript images, so you'll need to refer to the application's documentation. Many programs provide an Import choice on the File menu. Others provide an Insert menu or a Graphics menu. Note that if you use Encapsulated PostScript to embed the image, you will probably have to print the document on a PostScript printer in order for the image to be printed.

4.12.3 Using the MATLAB Print Command

The MATLAB print command permits you to send a model image to a printer, to the clipboard, or to a file in a variety of formats. The syntax of the print command is

```
print -smodel -ddevice filename
```

model is a MATLAB string containing the name of a currently open Simulink model. The model name is displayed in the title area of the model window. Note that filenames on your computer may be case-sensitive. If the model name contains spaces, enclose the name in single quotes:

```
print -s'Spring Mass System' -ddevice filename
```

TABLE 4.6: Device Codes for the `Print` Command

Device	Description
ps	PostScript
psc	Color PostScript
ps2	Level 2 PostScript
psc2	Level 2 color PostScript
eps	Encapsulated PostScript (must go to a file)
epsc	Color Encapsulated PostScript (must go to a file)
eps2	Encapsulated Level 2 PostScript (must go to a file)
epsc2	Color Encapsulated Level 2 PostScript (must go to a file)
win	Current printer
winc	Current printer, color
meta	Metafile format (saved to clipboard if no file name is supplied)
bitmap	Bitmap format (saved to clipboard if no file name is supplied)
setup	Same as selecting **File:Printer Setup** from the model window menu bar

If the model name contains a carriage return, that is, if it is shown in the window title in two lines, represent the carriage return as its ASCII code (13), enclose each line in single quotes ('), and the whole name in brackets ([ ]):

```
print -s['Damped' 13 'Spring Mass System'] -ddevice
filename
```

device is a MATLAB string that specifies the type of output device. Devices include printers, files, and the Windows clipboard. Listed in Table 4.6 are the available device types.

filename is a MATLAB string containing a valid file name, and it is an optional argument. If filename is specified, the output will be directed to the specified file rather than to the printer. If device is an Encapsulated PostScript format, and filename is not specified, filename will default to `Untitled.eps`. Note that the Encapsulated PostScript file will not contain a preview image, and therefore, the image will not be visible in some word-processing programs. The image will print correctly to a PostScript printer, however. It is possible to add a preview image using certain graphics programs, for example, GhostView.

■━━

EXAMPLE 4.3   Using the `print` command

In this example, we will, from the MATLAB prompt, print a model named `xy-demo.mdl` to the current default printer, then copy the model image to the clipboard in Windows Metafile format.

First, open model `xy-demo.mdl`. Next, at the MATLAB prompt, enter the following command:

```
print -s'xydemo'
```

The model will be printed. Next, enter the command

```
print -s'xydemo' -d'meta'
```

The model is now copied to the clipboard. It can be pasted into a word-processing document or spreadsheet.

─── ■

4.13 MODEL BUILDING SUMMARY

In Table 4.7, the model-building procedures discussed in Chapter 3 and Chapter 4 are summarized. The following terms are used:

Drag	Click and hold the left mouse button; drag object to new location.
Shift-click	Holding down the Shift key, click with the left mouse button.
Shift-drag	Holding down the Shift key, drag with the left mouse button.
Control-drag	Holding down the Control key, drag with the left mouse button (Windows only).
Right-drag	Drag using the right mouse button (Windows only).

TABLE 4.7: Summary of Model-building Operations

Operation	Method
Select object (block or signal line)	Click on the object with the left mouse button.
Select another object	Shift-click the additional object (or click the center mouse button on X Windows).
Select with bounding box	Click with the left mouse button at the location of one corner of the bounding box. Continuing to depress the mouse button, drag the bounding box to enclose the desired area.
Copy block from block library or another model	Select block, then drag it to the model window.
Flip block	Select block, then **Format:Flip Block**. Shortcut: Control-F.
Rotate block	Select block, then **Format:Rotate Block**. Shortcut: Control-R.
Resize block	Select block, then drag handle.
Add drop shadow	Select block, then **Format:Show Drop Shadow**.
Edit block name	Click on name.
Hide block name	Select name, then **Format:Hide Name**.
Flip block name	Select name, then **Format:Flip Name**. Shortcut: Drag name to new location.
Delete object	Select object, then **Edit:Clear**. Shortcut: Delete key.
Copy object to clipboard	Select object, then **Edit:Copy**. Shortcut: Control-C.
Cut object to clipboard	Select object, then **Edit:Cut**. Shortcut: Control-X.

TABLE 4.7: Summary of Model-building Operations (Cont)

Operation	Method
Paste from clipboard	**Edit:Paste**. Shortcut: Control-V.
Draw signal line	Drag from output port to input port.
Draw signal line in segments	Drag from output port to first bend. Release mouse button. Drag from first bend to second bend, and so on.
Branch from signal line	Control-drag from point of branch. Shortcut: Right-drag, starting from the point of branch.
Split signal line	Select line. Shift-drag the new vertex.
Move line segment	Select and drag segment.
Move line segment vertex	Select signal line and drag vertex marker.
Label signal line	Double-click line, then enter text.
Move signal line label	Drag the label to the desired location.
Copy signal line label	Control-drag the copy of the label to desired location. Shortcut: Right-drag to desired location.
Delete one occurrence of signal line label that has multiple occurrences	Shift-click label, then press Delete key.
Delete all occurrences of signal line label	Select label, then delete all characters in the label.
Propagate signal label	Label signal line onto which you want label propagated with single character <. Then choose **Edit:Update Diagram**.
Add annotation to model	Double-click at location of annotation, and type text.

4.14 SUMMARY

In this chapter, we described procedures for editing and annotating Simulink models. We also discussed configuring and printing a Simulink model. In the next two chapters, we will discuss using these procedures to model continuous, discrete, and hybrid systems.

REFERENCES

[1] Meirovitch, Leonard, *Introduction to Dynamics and Control* (New York: John Wiley & Sons, 1985) 16–17.

[2] Shampine, Lawrence F., and Reichelt, Mark W., "The MATLAB ODE Suite," The MathWorks, Inc. (Natick, Mass., 1996). This technical paper is available directly from The MathWorks. It provides a detailed discussion of the MATLAB differential equation solvers that are available from within Simulink.

[3] Hosea, M.E., and Shampine, Lawrence F., "Analysis and Implementation of TR-BDF2," *Applied Numerical Mathematics* 20, 1–2, (1996), 21–37.

5

Continuous Systems

In this chapter, we will discuss using Simulink to model continuous systems. We'll start with scalar linear systems. Then we'll model vector linear systems. Finally, we will use blocks from the Discontinuities, Math Operations, and User-Defined Functions block libraries to model nonlinear continuous systems.

5.1 INTRODUCTION

After studying Chapters 3 and 4, and experimenting a little, you should be comfortable with the mechanics of building and running Simulink models. In this chapter, we will explore using Simulink to model continuous systems.

As discussed in Chapter 2, a continuous system is a dynamical system that can be described using differential equations. Thus, most physical systems and processes are continuous. We'll start this chapter by modeling simple systems using blocks from the Continuous and Math Operations block libraries. Next, we'll show how to model more complex systems using vector signals. We'll also discuss using blocks from some of the other block libraries to model nonlinear continuous systems.

5.2 SCALAR LINEAR SYSTEMS

Scalar linear systems can be modeled using blocks in the Continuous and Math Operations block libraries. These blocks are easy to use, but the Integrator block has several important capabilities that merit elaboration. In this section, we will start with a detailed description of the Integrator block. Next, we will illustrate modeling scalar continuous systems with two examples.

5.2.1 Integrator Block

You can configure the Integrator block as a simple integrator or as a reset integrator. A reset integrator resets its output to the initial condition value when the reset signal triggers. You can also configure an Integrator block such that its output stays within preset limits. Additionally, you can set the integrator's initial output value in the Integrator block's dialog box, or configure the integrator to receive its initial output value through an additional input port.

Double-clicking an Integrator block opens the Integrator block dialog box, shown in Figure 5.1. To use a standard integrator, the only required input is **Initial condition**, which defaults to 0.

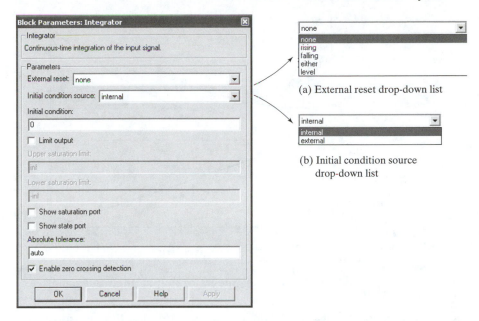

FIGURE 5.1: Integrator block dialog box

External reset is a drop-down list containing four choices: **none**, **rising**, **falling**, and **either**. Selecting **none** (the default) disables the external reset feature. If one of the other choices is selected, the Integrator block becomes a reset integrator block. If the reset choice is **rising**, the integrator output is reset to its initial condition value when the reset signal crosses (or departs rising, if it starts at zero) zero from below. If set to **falling**, the output is reset to the initial condition value when the reset signal crosses zero descending. If set to **either**, the output is reset to the initial condition value when the reset signal crosses zero from above or below. Figure 5.2 illustrates a simple model with the integrator **External reset** set to **falling**. Figure 5.3 shows the output and reset trajectories. When the reset signal crosses zero, the integrator output is reset to its initial value, and the simulation continues.

Setting **Initial condition source** to **external** adds an additional input port to the integrator. The value of this input will be used as the initial output of the integrator when the integration starts and when the integrator is reset, if an external reset is present.

Selecting **Limit output** causes the block to function as a limited integrator. The value of the output will be no greater than **Upper saturation limit** and no lower than

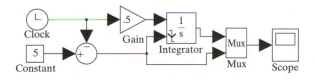

FIGURE 5.2: Model with reset integrator

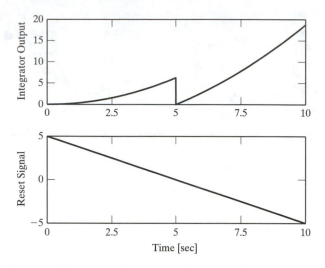

FIGURE 5.3: Reset integrator output and reset signals

Lower saturation limit. The default upper saturation limit is `inf`, which represents infinity, and the default lower saturation limit is `-inf`.

Selecting **Show saturation port** adds an additional output port that indicates the saturation status. The signal from this port will be −1 if the lower saturation limit is active, 1 if the upper saturation limit is active, and 0 if the output is between the saturation limits.

Selecting **Show state port** adds an additional output port to the block. This port will output the integrator state, which is the same as the integrator output. There are two situations in which the state port is needed. If the output of an Integrator block is fed back into the same block's reset or initial condition port, the state port signal must be used instead of the block output. You should also use a state port if you want to pass the output of an Integrator block in a conditionally executed subsystem (see Chapter 7 for details on conditionally executed subsystems) to another conditionally executed subsystem.

Absolute tolerance allows you to override the **Absolute tolerance** setting in the **Simulation:Parameters** dialog box for the output of a particular Integrator block. See Chapter 12 for a detailed explanation of **Absolute tolerance**. In a situation in which the absolute value of the output of an Integrator block is different by several orders of magnitude from other signals in a model, setting **Absolute tolerance** for that Integrator block appropriately can improve the accuracy of the simulation.

Shown in Figure 5.4 is an Integrator block with all possible ports active.

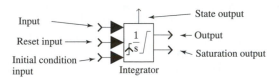

FIGURE 5.4: Integrator block with all options selected

EXAMPLE 5.1   Configuring an Integrator block

Consider the damped second-order system illustrated in Figure 5.5. Assume that

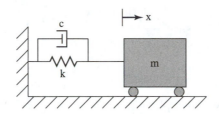

FIGURE 5.5: Damped second-order system

the damping coefficient $c = 1.0$ lb-sec/ft, the spring constant $k = 2$ lb/ft, and the cart mass $m = 5$ slugs. There is no input to this system. We will model the motion of the cart, assuming it is initially deflected 1 ft from the equilibrium position.

In order to model the system, it is necessary to write the equation of motion. Using the Newtonian approach, we note that there are two forces acting on the cart: the spring force and the damping force. The spring force is kx, and the damping force is $c\dot{x}$. The force due to the acceleration of the cart is $m\ddot{x}$. Because there are no externally applied forces, the sum of these three forces must be zero. Thus, we can write the equation of motion

$$m\ddot{x} + c\dot{x} + kx = 0$$

Because this is a second-order system, we will need two integrators to model its behavior. Open a new model window, then start building the model by dragging two integrators from the Continuous block library. Label them as shown, and connect the output of the Velocity integrator to the input of the Displacement integrator.

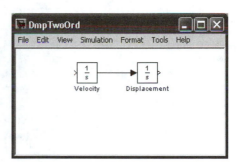

The output of the Velocity integrator is $\dot{x}$, so its input must be $\ddot{x}$. Rewrite the equation of motion to compute $\ddot{x}$ as a function of x and $\dot{x}$:

$$\ddot{x} = -\frac{k}{m}x - \frac{c}{m}\dot{x}$$

Substituting the values of the parameters,

$$\ddot{x} = -0.4x - 0.2\dot{x}$$

with $x(0) = 1, \dot{x}(0) = 0$.

Add a Sum block to compute $\ddot{x}$. The Sum block is configured with two minus signs ($| - -$).

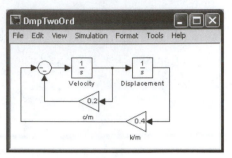

Add Gain blocks to compute the ratio of damping force to mass and the ratio of the spring force to mass. Draw signal lines as shown, and label the Gain blocks to show their functions more clearly.

Set the **Initial condition** field in the Velocity integrator's dialog box to 0 and the **Initial condition** field in the Displacement integrator's dialog box to 1.

Add a Scope block to display position.

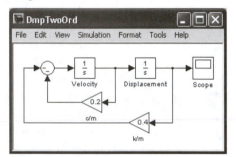

Choose **Simulation:Parameters** from the model window menu bar, and set **Stop time** to 50. Choose **Simulation:Start** from the model window menu bar. The Scope display should be as shown in Figure 5.6 after it is autoscaled.

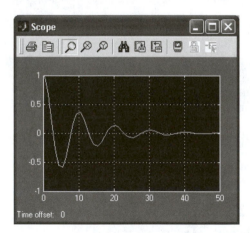

FIGURE 5.6: Damped second-order system response

5.2.2 Transfer Function Blocks

Simulink provides two blocks that implement transfer functions: the Transfer Fcn block and the Zero-Pole block. These blocks are equivalent, differing in the notation used to represent the transfer function. The Transfer Fcn block dialog box has two fields: **Numerator** and **Denominator**. **Numerator** contains the coefficients of the numerator of the transfer function in decreasing powers of s, and **Denominator** contains the coefficients of the denominator polynomial in decreasing powers of s. The Zero-Pole block dialog box has three fields: **Zeros**, **Poles**, and **Gain**. **Zeros** contains the zeros of the numerator of the transfer function, **Poles** contains the zeros of the denominator of the transfer function, and **Gain** scales the transfer function. Refer to the online help for detailed explanations of these blocks.

EXAMPLE 5.2   Using a Transfer Function block

Consider the spring-mass-damper system depicted in Figure 5.7. This system is the same as that in Example 5.1, with the addition of a forcing function F. Assume that the system is initially at its static equilibrium point $(x = 0, \dot{x} = 0)$, and that the forcing function is a step input of 1 lb. Ignoring friction, the equation of motion of this system is

$$m\ddot{x} + c\dot{x} + kx = F$$

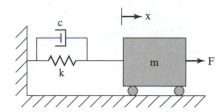

FIGURE 5.7: Forced second-order system

Taking the Laplace transform and ignoring initial conditions yields

$$m\,s^2 X(s) + c\,s X(s) + k X(s) = F(s)$$

The ratio of the Laplace transform of the output $[X(s)]$ to the Laplace transform of the input $[F(s)]$ is the transfer function $[G(s)]$,

$$G(s) = \frac{X(s)}{F(s)} = \frac{1/m}{s^2 + \frac{c}{m}s + \frac{k}{m}}$$

Shown in Figure 5.8 is a Simulink model of this system using the primitive linear blocks.

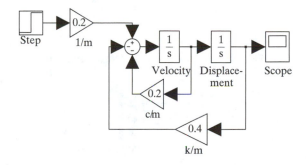

FIGURE 5.8: Forced second-order system with primitive blocks

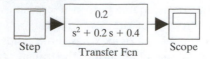

FIGURE 5.9: Forced second-order system using a Transfer Fcn block

In Figure 5.9, a Simulink model of the same system built using a single Transfer Fcn block is shown. The Transfer Fcn block dialog box field **Numerator** contains [0.2], and **Denominator** contains [1 0.2 0.4].

5.3 VECTOR LINEAR SYSTEMS

In the examples we examined so far, the signal lines carried scalar signals. It is often convenient to use vector signals, as they provide a more compact, easier-to-understand model. In this section, we will discuss the mechanics of using vector signals and then use them with the State-Space block.

5.3.1 Vector Signal Lines

You can combine several scalar signals to form a vector signal using the Mux (multiplexer) block from the Signals Routing block library. In Figure 5.10, we have combined three scalar signals to form a vector signal. Before you use a Mux block, you must configure it by setting the number of inputs. To make it easier to identify vector signal lines, choose **Format:Wide Vector Lines** from the model window menu bar. Choose **Format:Signal dimensions** to add a label showing the number of components of the vector line. The components of the vector signal are referred to as u(1), u(2), . . . , u(n), where n is the number of components. The top input to the Mux block is u(1); the bottom input is u(n).

The Demux block permits you to split a vector signal into a set of scalar signals. The Demux block must be configured for the correct number of outputs. Figure 5.11 illustrates splitting a vector signal into three scalar signals.

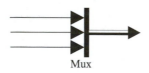

FIGURE 5.10: Forming a vector signal using a Mux block

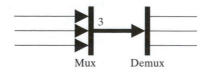

FIGURE 5.11: Splitting a vector signal into scalar signals

Most Simulink blocks will take vector inputs. The behavior of a block with a vector input will depend on the type of block and the block configuration parameters. Linear blocks with vector inputs produce vector outputs of the same dimension as the input. The configuration parameters for linear blocks with vector inputs must be of the same dimension as the input or must be scalar. In the case of scalar block parameters, Simulink automatically performs scalar expansion, which produces an implicit parameter vector of the same dimension as the input, and which has all elements set to the value of the block's scalar parameter. Consider the Simulink model fragment illustrated in Figure 5.12. The output of the Gain block is a three-element vector in which each element is the corresponding element of the input vector multiplied by 1. Now consider Figure 5.13. Notice that the gain is the vector [2,3,4], and the Gain block is configured for element-wise multiplication. In this case, the first element of the output is the first element of the input multiplied by 2, the second element of the output is the second element of the input multiplied by 3, and the third element of the output is the third element of the input multiplied by 4. The initial value of the output of an Integrator block behaves the same as the value of gain of a Gain block with respect to vector signals.

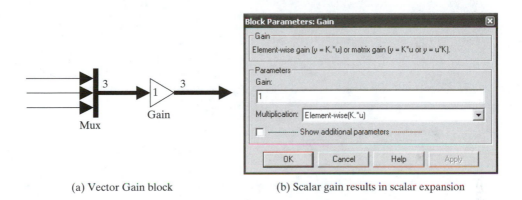

(a) Vector Gain block   (b) Scalar gain results in scalar expansion

FIGURE 5.12: Gain block with vector input signal

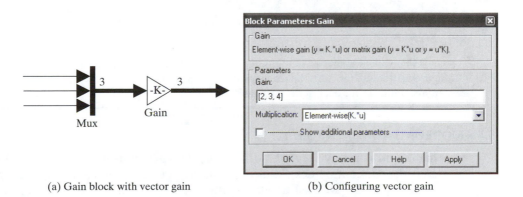

(a) Gain block with vector gain   (b) Configuring vector gain

FIGURE 5.13: Gain block with vector gain

Other blocks, such as the Fcn block (User-defined Functions block library), produce only scalar outputs, regardless of the input. The Block Browser describes the behavior of each block with respect to vector signals.

5.3.2 State-Space Block

The State-Space block implements a linear state-space model of a system or a portion of a system. The block dialog box has fields for each of the four linear state-space matrices (**A**, **B**, **C**, **D** as defined in Section 2.3.2) and a fifth field **Initial conditions**. Each field contains a MATLAB matrix.

■──

EXAMPLE 5.3   Impulse response of second-order system

Consider again the spring-mass system shown in Figure 5.7. In this example, we will model the impulse response of the system. Substituting in the model parameters, the equations of motion in state-space form are

$$\dot{x}_1 = x_2$$

$$\dot{x}_2 = -0.4\,x_1 - 0.2\,x_2 + 0.2\,\delta(t)$$

where $\delta(t)$ is the unit impulse function. The system matrix is

$$\mathbf{A} = \begin{bmatrix} 0 & 1 \\ -0.4 & -0.2 \end{bmatrix}$$

and the input matrix is

$$\mathbf{B} = \begin{bmatrix} 0 \\ 0.2 \end{bmatrix}$$

Define the output to be the block position,

$$\mathbf{C} = \begin{bmatrix} 1 & 0 \end{bmatrix}$$

There is no direct transmittance, so we can also set

$$\mathbf{D} = \begin{bmatrix} 0 \end{bmatrix}$$

We can approximate the unit impulse as a positive step function followed by a negative step:

$$\delta(t) \cong 100\,u(t) - 100\,u(t-1)$$

as was illustrated in Example 4.1.

Figure 5.14 shows a Simulink model of this formulation of the equations of motion. Figure 5.15 shows the State-Space block dialog box. Note that the State-Space block includes initial conditions for each state variable.

──■

5.4 MODELING NONLINEAR SYSTEMS

Simulink provides a variety of blocks for modeling nonlinear systems. These blocks are in the Discontinuities, Math Operations, and User-defined Functions block libraries. The behavior of nonlinear blocks with respect to vector inputs varies. Some blocks, such as the Relay, produce vector outputs of the same dimension as the input. Other blocks produce only scalar outputs, or scalar or vector outputs,

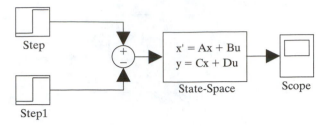

FIGURE 5.14: State-Space model of spring–mass system

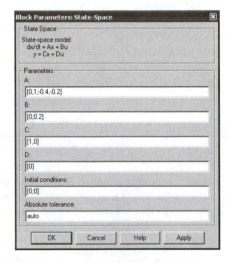

FIGURE 5.15: State-Space dialog box

depending on the dimensionality of the inputs. Consult the online help for details on a particular block.

EXAMPLE 5.4   Rocket-powered cart model

To illustrate the use of several nonlinear blocks, we will model the motion of the rocket-powered cart shown in Figure 5.16. The cart is powered by two opposing rocket motors. The controller fires the left motor if the sum of cart velocity and displacement is negative, and it fires the right motor if the sum of cart velocity and displacement is positive. The objective of the controller is to bring to cart to rest at the origin. This type of control is sometimes called bang-bang control. (This is actually a simple example of a very powerful technique called sliding mode control, explained by Khalil [1].)

Assume that the cart mass is 5 slugs, and that the motor force F is 1 lb.

We begin by writing the equation of motion of the cart:

$$m\ddot{x} = F$$

This is a second-order system, so two Integrator blocks are needed to solve for the cart position.

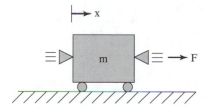

FIGURE 5.16: Rocket-powered cart

Open a new model window, and copy two Integrator blocks from the Continuous block library. Label the blocks as shown.

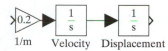

The input to the first integrator block is acceleration. Solving the equation of motion for acceleration,

$$\ddot{x} = \frac{F}{m}$$

Add a Gain block to multiply the rocket motor force by $1/m$. The input to this Gain block will be the motor force F.

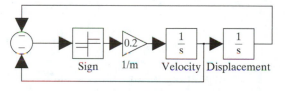

The motor force F is 1.0 if the sum of velocity and displacement is negative, and -1.0 if the sum is positive. We can build a suitable bang-bang controller using a Sum block and a Sign block from the Math block library. The output of a Sign block is 1.0 if its input is positive, -1.0 if its input is negative, and 0.0 if its input is 0.

Add the Sum block and Sign block, and connect them as shown. The Sum block is configured with two minus signs (- -) because the motor force is to be opposite in sign to the sum of velocity and displacement.

To display the results of the simulation of the model, we will use an XY Graph block to draw a *phase portrait* as the simulation progresses. A phase portrait is a plot of velocity versus displacement.

Drag an XY Graph block from the Sinks block library. Connect the output of the displacement Integrator to the X input (the upper input port) and the output of the velocity Integrator to the Y (lower) input port.

To display simulation time as the simulation progresses, add a Clock block from the Sources block library, and connect it to a Display block from the Sinks block library.

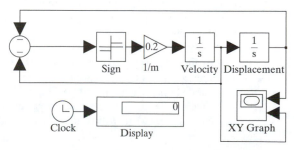

We don't know how long it will take the solution to reach the origin, so for convenience, let's add logic to the model to cause the simulation to automatically stop when the objective is reached. Specifically, we will add logic that causes the simulation to stop when the sum of the absolute values of velocity and displacement falls below a threshold value of 0.01. We can accomplish this task with a Stop Simulation block from the Sinks block library. This block forces Simulink to stop the simulation when the block's input is nonzero. We wish for the input to the Stop Simulation block to be zero until

$$|x| + |\dot{x}| \leq 0.01$$

Copy two Abs blocks, a Sum block, and a Relational Operator block from the Math Operations block library. Double-click the Relational Operator block, and choose <= from the Operator drop-down list. Add a Constant block from the Sources block library, and set its value to 0.01. Finally, drag a Stop Simulation block from the Sinks block library, and connect the signal lines as shown.

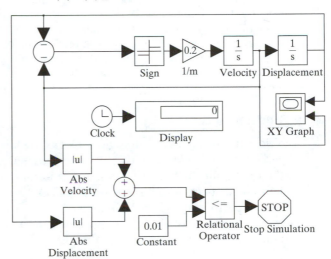

We will assume that the cart is initially at rest, displaced 1 ft to the right. So, set the **Initial condition** for the velocity Integrator block to 0, and that for the displacement Integrator block to 1.

The Sign block switches instantaneously when its input changes sign. However, any physically realizable switch takes a finite amount of time to change state. To model that

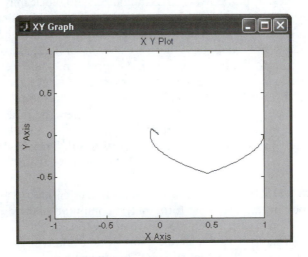

FIGURE 5.17: Cart phase portrait

behavior, we can use a fixed step size solver. Let's assume that the switch time is 0.05 seconds. Choose **Simulation:Parameters** from the model window menu bar. Set the Solver options **Type** fields to Fixed-step and ODE5 (Dormand-Prince). Then, set the **Fixed step size** to 0.05. Set **Stop time** to 200.

Now, we are ready to run the simulation. Choose **Simulation:Start** from the model window menu bar. The simulation will execute and will stop when the cart is at rest at the origin. The XY Graph should appear as shown in Figure 5.17.

If you look closely at the phase portrait, you will see that as the phase trajectory approaches the origin, it oscillates about the line $\dot{x} = -x$. This phenomenon is called *chatter* (discussed in detail by Khalil [1]). Try replacing the Sign block with a Saturation block with **Upper limit** set to 0.05 and **Lower limit** set to -0.05. Also try replacing the Sign block with a Dead Zone block, with **Start of dead zone** set to -0.05 and **End of dead zone** set to 0.05. When you use the Saturation block or Dead Zone block instead of the Sign block, you can use a variable-step solver, because none of the blocks changes instantaneously from −1 to +1.

5.4.1 Function Blocks

Two particularly useful nonlinear blocks are the Fcn (C function) block and the MATLAB Fcn block. Both blocks perform specified mathematical operations on the input, but the blocks have some important differences. Simulink evaluates the Fcn block much faster than the MATLAB Fcn block. However, the Fcn block cannot perform matrix computations, use MATLAB functions, or produce a vector output. Because of the speed advantage, it is always better to use the Fcn block, unless the special capabilities of the MATLAB Fcn block are needed.

Fcn Block.   The Fcn block is shown in Figure 5.18. The dialog box contains a single field that contains an expression in the C language syntax. The expression operates on the elements of the block input vector, referring to the elements as u[n], where n is the desired element. If the input is a scalar, the input is referred to as u[1]. Note the use of square brackets ([and]) as in the C language, rather than parentheses, as in the MATLAB language. The block may also use as a parameter any variable currently (at the time of execution of the simulation) defined in the MATLAB workspace. If the workspace variable is a scalar, it may be referred to by its name. For example, if there is a scalar workspace variable a and two-element vector input to the Fcn block, a valid expression would be: a*(sin(u[1])+u[2]). If the workspace variable is a vector or matrix, it is referred to using the appropriate MATLAB (not C) syntax: A(1), or A(2,3) (note the parentheses here). The Fcn block can perform all the standard scalar mathematical functions such as sin, abs, atan, C syntax relational operations (==, !=, >, <, >=, <=), and C syntax logical operations [&& (logical AND), || (logical OR)]. The Fcn block produces a scalar output.

MATLAB Fcn Block.   The MATLAB Fcn block is more powerful than the Fcn block in that it can perform matrix computations and produce vector outputs. However, it is also much slower than the Fcn block, and thus, should be used only in situations in which the Fcn block cannot be used. The MATLAB Fcn block is illustrated in Figure 5.19. The dialog box has three fields. The first may contain any valid MATLAB expression, in standard MATLAB syntax. The value

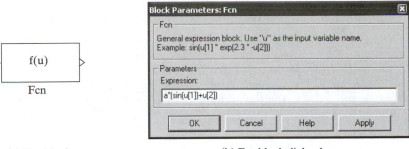

(a) Fcn block

(b) Fcn block dialog box

FIGURE 5.18: Fcn block

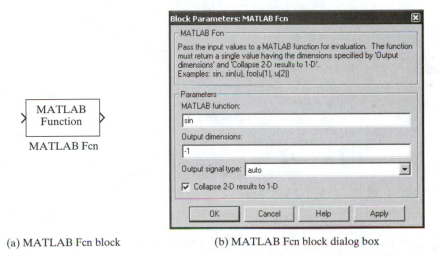

(a) MATLAB Fcn block

(b) MATLAB Fcn block dialog box

FIGURE 5.19: MATLAB Fcn block

of the expression is the block output. The n–th element of the block input vector is referred to as u(n), similar to the Fcn block. If the function field contains a MATLAB function with no arguments [as in Figure 5.19(b)], the operation is performed on all elements of the input. The second input field specifies the dimension of the output vector. Enter -1 if the output is to be the same width as the input. No matter whether Output width is specified explicitly or allowed to default to the input width, the vector dimension of the result of the expression in the MATLAB function field must be the same as the value specified in Output width. The third field specifies the output data type. There is also a checkbox that forces vector output to be one-dimensional.

The final continuous example will illustrate the use of nonlinear blocks to build a simple model of a car and proportional gain cruise control. The model will include aerodynamic drag, the gravity force due to climbing hills, and the wind.

■
EXAMPLE 5.5   Automobile proportional speed control

Consider the automobile traveling on a straight, hilly road shown in Figure 5.20. There are three forces acting on the automobile: the forward thrust produced by the engine and transmitted through the tires (or the braking force if negative) (F_e), the aerodynamic force (including wind) (F_w), and the tangential component of gravity as the automobile climbs and descends hills (F_h). Applying Newton's second law, the equation of motion of the automobile can be written

$$m\ddot{x} = F_e - F_w - F_h$$

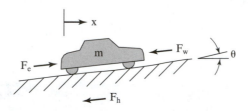

FIGURE 5.20: Automobile on a hilly road

where m represents the mass of the automobile and x the distance traveled. F_e must have upper and lower bounds. The upper bound is the maximum force that the engine can transmit through the wheels to the road, and the lower bound is the maximum braking force. We will assume that $-2000 \leq F_e \leq 1000$, with units of lb, and that the mass is 100 slugs.

The aerodynamic force is the product of the drag coefficient (C_D), the automobile's frontal area (A), and the dynamic pressure (P), where

$$P = \frac{\rho V^2}{2}$$

and ρ represents the air density and V the sum of the automobile speed and wind speed (V_w). Assume that

$$\frac{C_D A \rho}{2} = 0.001$$

and that the wind speed varies sinusoidally with time according to the rule

$$V_w = 20 \sin(0.01\,t)$$

so that the aerodynamic force can be approximated by

$$F_w = 0.01 \, (\dot{x} + 20 \, \sin(0.01\,t))^2$$

Next, assume that the road angle varies sinusoidally with distance, according to the rule

$$\theta = 0.0093 \sin(0.0001x)$$

Then the hill force is

$$F_h = 30 \sin(0.0001x)$$

We will control the automobile speed using the simple proportional control law

$$F_c = K_e \, (\dot{x}_{desired} - \dot{x})$$

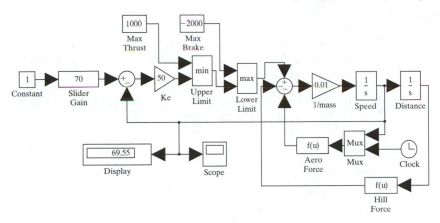

FIGURE 5.21: Automobile model with proportional speed control

Here, F_x is the commanded engine (or braking) force, $\dot{x}_{desired}$ is the commanded speed (ft/sec), and K_e is the feedback gain. Thus, the command engine force is proportional to the speed error. The actual engine force (F_e) is, as stated earlier, bounded from above by the maximum engine thrust from below by the maximum braking force. We choose $K_e = 50$.

A Simulink model of this system is shown in Figure 5.21. We will simulate the motion of the car for 1000 sec.

The input to the proportional controller is the desired automobile speed in ft/sec. This is implemented with a Slider Gain block (from the Math block library) with a constant input. Double-click the Slider Gain block to open a slider window, which allows you to vary the desired speed while the simulation runs.

The proportional controller consists of a Sum block that computes the speed error (the difference between the commanded speed and the actual speed) and a Gain block.

The upper and lower limits on engine force are imposed using MinMax blocks. The constant blocks labeled Max thrust and Max Brake, together with the Min and Max blocks, are used here to illustrate the use of those blocks. This part of the model could be replaced with a Saturation block from the Discontinuities block library. (Why don't you try that?)

The nonlinear hill and aerodynamic forces are computed by Fcn blocks. The **Expression** field of the block dialog box for the Fcn block labeled Aero Force contains 0.001*(u[1]+20*sin(0.01*u[2]))^2. The **Expression** field for the Fcn block labeled Hill Force contains 30*sin(0.0001*u[1]).

A Display block serves as a speedometer (which indicates ft/sec), and the speed is plotted using a Scope block.

This model is a good example of a slightly stiff system. To observe the effect of this stiffness, run the model using solver ODE45, then repeat the simulation using ODE15S. See Chapter 13 for a discussion of stiff systems.

5.5 SUMMARY

In this chapter, we discussed using Simulink to model continuous systems. We started with scalar, linear, time-invariant systems, and progressed to vector linear systems using the state-space concept. Finally, we discussed modeling nonlinear continuous systems.

REFERENCES

[1] Khalil, Hassan K., *Nonlinear Systems*, 2nd ed. (Upper Saddle River, N.J.: Prentice Hall, 1996). This is an excellent text on the analysis and design of nonlinear systems.

[2] Kuo, Benjamin C., *Automatic Control Systems* (Englewood Cliffs, N.J.: Prentice Hall, 1995), 226–230. This is a comprehensive text covering all the basics of control system analysis and design.

[3] Lewis, Paul H., and Yang, Charles, *Basic Control Systems Engineering* (Upper Saddle River, N.J.: Prentice Hall, 1997). This text provides an introduction to control systems analysis and design and includes a brief introduction to Simulink. Simulink is used in many of the examples.

[4] Ogata, Katsuhiko, *Modern Control Engineering* (Englewood Cliffs, N.J.: Prentice Hall, 1990). This book presents a thorough coverage of the standard techniques for the analysis and design of controls for continuous systems.

[5] Scheinerman, Edward C., *Invitation to Dynamical Systems* (Upper Saddle River, N.J.: Prentice Hall, 1996). This text provides a good introduction to nonlinear systems.

[6] Shahian, Bahram, and Hassul, Michael, *Control System Design Using MATLAB* (Englewood Cliffs, N.J.: Prentice Hall, 1993). This book provides an introduction to MATLAB programming, and uses MATLAB to solve many of the standard problems in classical control and modern control theory.

[7] Strum, Robert D., and Kirk, Donald E., *Contemporary Linear Systems Using MATLAB* (Boston, Mass.: PWS Publishing Co., 1994). Chapter 2 provides a nice introduction to continuous systems and the state-space concept.

[8] Vidyasagar, M., *Nonlinear Systems Analysis*, 2nd ed. (Englewood Cliffs, N.J.: Prentice Hall, 1993). This book presents detailed coverage of the analysis of nonlinear systems.

6

Discrete-Time Systems

In this chapter we will discuss using Simulink to model discrete-time systems. First, we'll model discrete-time systems using blocks from the Discrete, Math Operations, User-Defined Functions, and Look-Up Tables block libraries. Then, we'll discuss modeling hybrid systems, which include continuous and discrete components.

6.1 INTRODUCTION

Simulink provides extensive capabilities for modeling discrete-time systems. The Discrete block library contains simple discrete blocks such as Zero- and First-Order Holds, Discrete integrators, and the Unit Delay. The Math Operations block library contains additional blocks that are useful in discrete-time systems, such as logical operators and a Combinatorial Logic (truth table) block. Additionally, a number of blocks in the Math Operations (such as Sum and Gain) and User-Defined Functions (for example, Fcn) block libraries have the same purpose when used with continuous or discrete-time models. Discrete-time models may be single-rate, or they may be multirate, including offset sample times. Simulink will optionally color-code the signal lines in multirate models such that the signal line color indicates the sample period.

6.1.1 Scalar Linear Discrete-Time Systems

Modeling scalar linear discrete-time systems is similar to modeling continuous systems. Discrete-time models can use the Gain and Sum blocks from the Math block library. These blocks behave the same in discrete systems as they do in continuous systems. The Discrete block library contains the discrete analogs to the continuous Integrator and Transfer Fcn blocks. We will discuss these blocks next, and show how they may be used.

Each discrete block is assumed to have a sampler at its input and a zero-order hold at its output. Discrete blocks have the additional configuration parameter **sample time**. **Sample time** is a scalar interval between samples or a two-element vector consisting of the interval between samples and an offset or time skew. For example, if the block is to have a sample interval of 1.5 seconds, and no offset, **Sample time** would be set to 1.5, and samples would be taken at $t = 0, 1.5, 3.0$ seconds. If the sample time is to be 1.5 seconds, with a 0.5 second offset, the Sample time would be set to [1.5, 0.5], and samples would be taken at $t = 0.5, 2.0, 3.5$ seconds.

All of the differential equation solvers listed in the **Simulation:Parameters Solver options** are compatible with discrete systems. A special solver, named "discrete (no continuous states)," is the best choice for purely discrete systems, because this solver is optimized for these systems.

6.1.2 Unit Delay

As discussed in Chapter 2, the Unit Delay is the fundamental discrete-time block. The output of the unit delay block is the input at the previous sample time. The unit delay represents the difference equation

$$y(k) = x(k-1)$$

where y is the output sequence, and x is the input sequence. The Unit Delay block dialog box has two fields. The first field is **Initial condition**. This field contains the value of the block output at the start of the simulation. The second field is **Sample time**.

EXAMPLE 6.1   Loan amortization model

In this example, we will build a Simulink model for the amortization of an automobile loan, as discussed in Example 2.6. At the end of each month, the loan balance $b(k)$ is the sum of the balance at the beginning of the month $[b(k-1)]$ and the interest for the month $[i\,b(k)]$, less the end-of-month payment $p(k)$. Thus, the balance at the end of month k is

$$b(k) = r\,b(k-1) - p(k)$$

where $r = i + 1$, and i is the monthly interest rate.

Assume that the initial loan balance is \$15,000, the interest rate is 1% per month (12% annual interest), and the monthly payment is \$200. Compute the loan balance after 100 payments.

Figure 6.1 shows a Simulink model of this system. The Unit Delay block computes $b(k-1)$. The Unit Delay block **Initial condition** is the initial loan balance (15,000). The Unit Delay block **Sample time** is set to 1. Set solver type to `Fixed-step, discrete (no continuous states)`, and let **Start time** be 0 and **Stop time** be 100. After running the simulation, the Display block shows the ending balance.

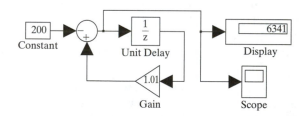

FIGURE 6.1: Block diagram of loan amortization

6.1.3 Discrete-Time Integrator

The Simulink Discrete-Time Integrator block can be configured to model each of the three varieties of discrete-time integrators discussed in Section 2.4.4. The

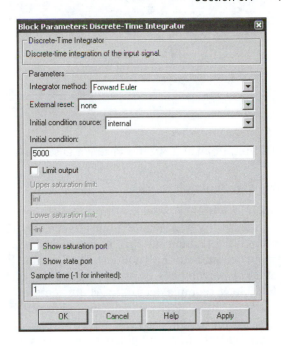

FIGURE 6.2: Discrete-Time Integrator block dialog box

Discrete-Time Integrator block dialog box is shown in Figure 6.2. The fields in this dialog box are the same as the fields for the Integrator dialog box, with two additions. For a detailed discussion of the Integrator block dialog box fields, refer to Section 5.2.1. The two additional fields are **Integrator method** and **Sample time**.

   Integrator method is one of three choices: Forward Euler, Backward Euler, and Trapezoidal. Because the Discrete integrator is a discrete block, it has a sample and zero-order hold at its input. Thus, at each time step, a Discrete integrator block has access to only two values of $u(t)$: $u(kT)$ at its input and $u((k-1)T)$ via a built-in unit delay.

EXAMPLE 6.2   Loan amortization model

Consider the automobile loan problem of Example 6.1. The loan balance at the end of the month can be computed as

$$b(k) = b(k-1) + \int_{T(k-1)}^{T(k)} [i\,b(k-1) - p]\,dt$$

   Examining the integrand, we can see that this problem can be solved using forward Euler integration. Figure 6.3 shows the Simulink model revised to use a Discrete-Time Integrator block with **Integrator method** set to Forward Euler. The Discrete-Time Integrator block **Initial condition** is set to 15,000 and Sample time to 1. Note that, in this case, the feedback gain is set to 0.01. Also note that the simulation Stop time must be set to 101, because the forward Euler integrator evaluates the rate at the beginning of a period. Try changing the integrator to Backward Euler and Trapezoidal.

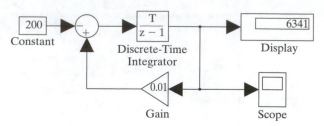

FIGURE 6.3: Loan amortization using a Discrete-Time Integrator block

6.1.4 Discrete Transfer Function Blocks

The Discrete block library provides three blocks that implement discrete transfer functions: Discrete Filter, Discrete Transfer Fcn, and Discrete Zero-Pole. These blocks are equivalent, differing only in the definitions of the coefficients for the numerator and denominator polynomials. The Discrete Filter block requires vectors of coefficients of polynomials of ascending powers of z^{-1}, whereas the Discrete Transfer Fcn block requires vectors of coefficients of polynomials of descending powers of z. Thus, these blocks are identical, differing only in the way the transfer function is displayed on the block icon. The Discrete Zero-Pole block requires vectors of zeros (numerator) and poles (denominator) of the transfer function and a gain that scales the transfer function.

■

EXAMPLE 6.3   Discrete Butterworth filter

Control systems frequently contain low-pass filters to remove high-frequency noise from input signals. The MATLAB Signal Processing Toolbox provides a variety of filter design algorithms. You can include a filter designed using MATLAB in a Simulink model using a Discrete Transfer Fcn block.

Suppose we are designing a control system with a noisy sinusoidal input, and we need to filter out the noise. The sampling period is 0.1 second, and we wish to remove signals with a frequency higher than 1.0 Hz. Design a fourth-order Butterworth filter using the MATLAB command `butter` as shown in Listing 6.1.

Listing 6.1: Designing a Butterworth filter

```
>> [B,A]=butter(4,0.2)
B =
    0.0048    0.0193    0.0289    0.0193    0.0048
A =
    1.0000   -2.3695    2.3140   -1.0547    0.1874
```

The first argument to `butter` is the filter order, and the second argument is the cutoff frequency ω_m, where $0 < \omega_m < 1$ (and where 1 corresponds to one-half the sample rate). So here, $\omega_m = 1\,\text{Hz}/5\,\text{Hz} = 0.2$. (See Orfanidis [4] to learn about digital filters.)

We can test the filter with the Simulink model shown in Figure 6.4. The top Sine Wave block is configured to a frequency of 0.5 rad/sec and an amplitude of 1. The lower Sine Wave block is configured to a frequency of 10 rad/sec and an amplitude of 0.4 and represents the unwanted high-frequency noise. The Discrete Transfer Fcn block field **Numerator** contains the vector [0.0048 0.0193 0.0289 0.0193 0.0048] and **Denominator** contains [1.0000 −2.3695 2.3140 −1.0547 0.1874]. **Sample time** for the Sine wave blocks and Discrete Transfer Fcn block contains 0.1. The model is configured to use the discrete solver, because there are no continuous states. Stop time is set to 20. The Mux block creates a vector signal containing the unfiltered input and the filtered output of the Discrete Transfer Fcn block. Scope block field **Save data to workspace** is checked. After running the simulation, the MATLAB commands shown in Listing 6.2 are used to plot the filtered and unfiltered output signal, resulting in the plots shown in Figure 6.5.

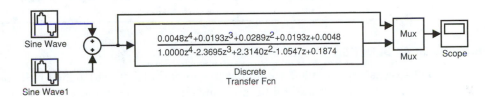

FIGURE 6.4: Simulink model to test filter

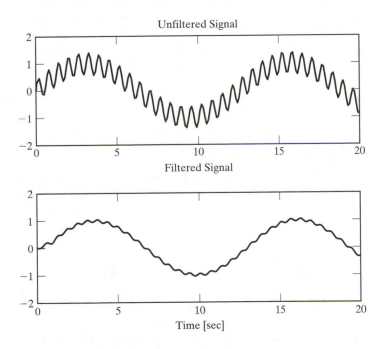

FIGURE 6.5: Filtering a noisy signal

Listing 6.2: MATLAB commands to plot filtered signal

```
>> t = ScopeData(:,1);
>> y_raw = ScopeData(:,2);
>> y_filt = ScopeData(:,3);
>> subplot(2,1,1)
>> plot(t,y_raw);
>> title('Unfiltered Signal')
>> grid
>> subplot(2,1,2)
>> plot(t,y_filt);
>> title('Filtered Signal')
>> xlabel('Time (sec)')
>> grid
```

6.2   LOGICAL BLOCKS

The Logical Operator and Combinatorial Logic blocks found in the Math operations block library are also useful in modeling discrete-time systems. These are discrete rather than discrete-time blocks, because they do not have samplers at their inputs or zero-order holds at their outputs. In fact, these blocks can be used in continuous models, although they are more commonly used in discrete-time models. The Logical Operator block can be configured to perform any of the following logical operations: AND, OR, NAND, NOR, XOR, NOT. The Logical Operator block can be configured to accept any number of inputs. If the input signals are vectors, each must be of the same size, and the output will also be a vector.

The default data type for logical blocks is Boolean. Therefore, Simulink will produce an error message when standard double-precision signals are input to a logical block. To suppress these error messages, open the **Simulation:Parameters** dialog box, and select the **Advanced** page. Set item `Boolean logic signals` to **Off**. This feature should be enabled if you are using Boolean data and wish to force logical blocks to use Boolean data.

The Combinatorial Logic block implements a truth table. The input to the Combinatorial Logic block is a vector of n elements. The truth table must have $2n$ rows, arranged such that the values of a row's inputs provide an index into the table. The first element in the input vector is the left-most column in this binary index. Thus, if there is a single input, there will be two rows: the first corresponding to an input of 0 and the second corresponding to an input of 1 (or any nonzero number). If there are two inputs, there must be four rows in the table. The block parameter, Truth table, consists of one or more column vectors, in which each row represents the block output corresponding to the row's index. Each column in the Truth table produces a different output, and thus, a single Combinatorial Logic block can implement multiple logical operations. The block outputs do not have to be 0 or 1; any number is acceptable.

EXAMPLE 6.4   Combinatorial Logic blocks

We will build Combinatorial Logic blocks that implement the three following expressions, and we will compare these three blocks with equivalent models built using Logical Operator blocks.

Expression	Equivalent
$\overline{a}$	NOT a
ab	a AND b
$ab + bc$	$(a$ AND $b)$ OR $(b$ AND $c)$

The truth table for $\overline{a}$ is shown in Figure 6.6(a). The corresponding Truth table parameter is the output column, entered as [1;0]. A Simulink model that implements this expression using Logical Operator blocks and a Combinatorial Logic block is shown in Figure 6.6(b).

a	**Output**
0	1
1	0

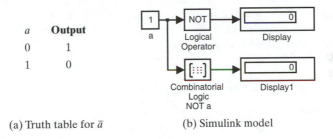

(a) Truth table for $\overline{a}$ (b) Simulink model

FIGURE 6.6: NOT expression

The truth table for ab is shown in Figure 6.7(a). Note that we chose a to be the left-most column, thus the input vector to the block must be $[a, b]$. The block **Truth table** parameter is the same as the output column, [0;0;0;1]. Figure 6.7(b) shows a Simulink model that implements ab using both Logical Operator blocks and a Combinatorial Logic block. The input vector is composed using a Mux block such that the components are in the order $[a, b]$.

a	b	**Output**
0	0	0
0	1	0
1	0	0
1	1	1

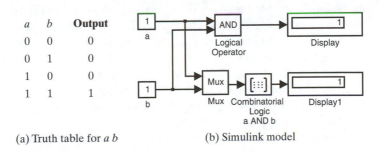

(a) Truth table for $a\,b$ (b) Simulink model

FIGURE 6.7: AND expression

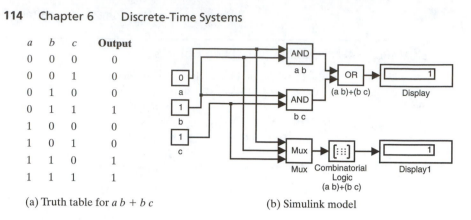

a	b	c	Output
0	0	0	0
0	0	1	0
0	1	0	0
0	1	1	1
1	0	0	0
1	0	1	0
1	1	0	1
1	1	1	1

(a) Truth table for $a\,b + b\,c$ (b) Simulink model

FIGURE 6.8: Model of the expression $ab + bc$

The final combinatorial logic expression, $ab + bc$, has the truth table shown in Figure 6.8(a). Examining the output column, the **Truth table** parameter is [0;0;0;1;0;0; 1;1]. Figure 6.8(b) implements the expression using both Logical Operator blocks and a Combinatorial Logic block.

6.3 VECTOR DISCRETE-TIME SYSTEMS

The Discrete block library provides a Discrete State-Space block as defined in Section 2.5. The use of this block is completely analogous to the State-Space block discussed in 5.3.2.

6.4 MULTIRATE DISCRETE-TIME SYSTEMS

Many discrete-time systems involve subsystems that operate at different rates and with phase differences. For example, a typical computer has numerous discrete subsystems, including the central processing unit, serial and parallel interface controllers, disk drive and video controllers, and input devices such as the keyboard and mouse. Models of communications systems and transaction processes may also consist of subsystems that operate at different rates. Modeling multirate discrete-time systems with Simulink is similar to modeling single-rate discrete-time systems. However, multirate systems require careful attention to sample times and offsets.

The Simulink sample time colors capability provides help in keeping track of sample times. This capability will automatically color code blocks and signal lines corresponding to up to five different values of sample time. To activate this feature, choose **Format:Sample Time Colors**. If you change the model after activating sample time colors, choose **Edit:Update Diagram** to update the colors. Table 6.1 lists the sample time colors and their meanings.

Inherently continuous blocks (such as integrators) and inherently discrete blocks (such as unit delays) are assigned colors using the definitions in Table 6.1. Signal lines and blocks that are neither inherently continuous nor discrete (such as Gain blocks) are assigned colors based on the sample times of their inputs. Thus, if a signal line is carrying the output of an Integrator, it will be black; if it is carrying the

TABLE 6.1: Sample Time Colors

Color	Meaning
Black	Continuous blocks
Magenta	Constant blocks (used with Real-Time Workshop)
Yellow	Hybrid (groups of blocks with varying sample times or mixed continuous and discrete elements)
Red	Fastest discrete sample time
Green	Second fastest discrete sample time
Blue	Third fastest discrete sample time
Light blue	Fourth fastest discrete sample time
Dark green	Fifth fastest discrete sample time
Orange	Sixth fastest discrete sample time
Cyan	Triggered sample time (used with triggered subsystems, see Chapter 7)
Gray	Updates at major time steps only

output of a Unit Delay, it will be colored according to the sample time of the Unit Delay. If the input to a block (such as a Sum block) consists of several signals, such that the sample times of all signals are integer multiples of the sample time of the fastest signal, the block is colored corresponding to the fastest signal. If the sample times are not integer multiples of the fastest signal, the block is colored black.

Only the six fastest discrete sample times are assigned unique colors. If there are more than six sample times, all blocks sampled slower than the sixth-fastest sample time are colored yellow.

Note that Simulink does not provide a mechanism to help identify phase relationships among blocks. If there are portions of a model with the same sample time but different values for sample offset (phase), you must keep track of the phase manually.

■

EXAMPLE 6.5   Multirate discrete system

Control systems for discrete processes frequently operate at a lower frequency than the update frequency of the process, usually due to limitations of computer speed. Additionally, display systems are usually updated at a frequency sufficiently low such that the display is readable. As a simple example of a multirate system, suppose that some process to be controlled behaves according to the following discrete state-space equations:

$$x_1(k+1) = x_1(k) + 0.1x_2(k)$$

$$x_2(k+1) = -0.05 \sin x_1(k) + 0.094x_2(k) + u(k)$$

where $u(k)$ is the input. The process is assumed to have a sample time of 0.1 sec. We will control the process using proportional control with a sample time of 0.25 sec and update the display every 0.5 sec. A Simulink model of this system appears in Figure 6.9. The model is annotated to indicate the colors of the signal lines and blocks.

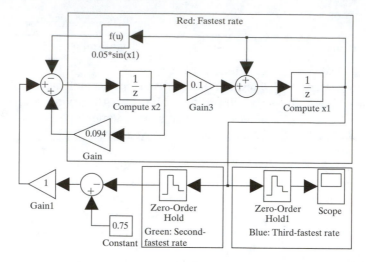

FIGURE 6.9: Multirate system

The blocks in the section of the model with the red signal lines are configured with a sample time of 0.1 sec. This section represents the process dynamics.

The section with green signal lines models the controller. This is a proportional controller. The Zero-Order Hold block causes the controller to update at its sample interval of 0.25 sec. The controller produces an output signal that is proportional to the difference between the setpoint (here 0.75) and the value of the input to the Zero-Order Hold (x_1) at the most recent sample time.

The section with the blue signal lines models the display device, here a Scope block. The Zero-Order Hold in this section is configured with a sample time of 0.5 sec.

Running the simulation results in the trajectory plotted in Figure 6.10.

6.5 HYBRID SYSTEMS

As discussed in Section 2.7, hybrid systems consist of discrete and continuous components. Modeling hybrid systems in Simulink is straightforward, as illustrated in the following example.

EXAMPLE 6.6 Discrete PID automobile speed control

To illustrate the construction of a hybrid model, let's replace the continuous controller in Example 5.5 with a discrete proportional-integral-derivative (PID) controller with a sample time of 0.5 sec.

Figure 6.11 illustrates a continuous PID controller. The controller consists of three sections, each of which operates on the difference (v) between the plant output and the commanded value of the plant output. The proportional section produces a signal proportional to the difference between the commanded value of the system output and the actual value. Thus, the output of the proportional section is

$$u_p = K_p v$$

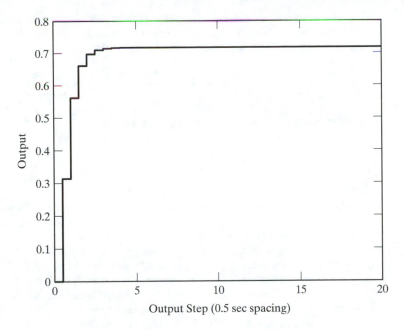

FIGURE 6.10: Multirate system output

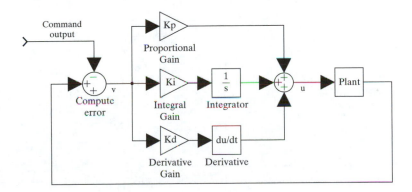

FIGURE 6.11: Continuous PID controller

The integral section of a PID controller is intended to remove the steady-state error. This component produces an output that is proportional to the time integral of the error signal:

$$u_i = K_i \int_0^t v \, dt$$

One problem with the integral section is that if the plant's response to changes in the input signal (u) is relatively slow, u_i can grow rapidly. This phenomenon is called *wind-up*. Wind-up can be avoided by placing upper and lower bounds on the value of u_i.

The derivative section of a PID controller provides damping. Its output is proportional to the rate of change of v:

$$u_d = K_d \, \dot{v}$$

A discrete PID controller replaces the integral section with a discrete integrator and the derivative section with a discrete approximation to a derivative block. For a detailed discussion of discrete PID controllers, refer to Ogata [3]. A first-order numerical derivative approximation is

$$u_d(k) \approx \frac{v(k) - v(k-1)}{T}$$

The transfer function of this derivative approximation is

$$\frac{U_d(z)}{V(z)} = \frac{K_d}{T}\left(\frac{z-1}{z}\right)$$

Figure 6.12 shows a Simulink model of an automobile using a discrete PID controller with $K_p = 50$, $K_i = 0.75$, and $K_d = 75$. The model is identical to the model in Example 5.5, except for the controller. The proportional part of the controller consists of a Zero-Order hold and a proportional Gain block. The proportional gain (50) in this controller is the same as the proportional gain in the continuous proportional controller in Example 5.5. The integral part of the PID controller consists of a Discrete-Time Integrator block and a Gain block (set to 0.75). In the Discrete-Time Integrator block, we selected **Limit output** and set the saturation limits to ± 100 to control wind-up. The derivative part of the controller is constructed using a Discrete Transfer Fcn block and a Gain block (set to 75). In this example, the simulation was configured to run for 1000 sec, and the Slider Gain was set to command a speed of 80 ft/sec. The Scope display is shown in Figure 6.13.

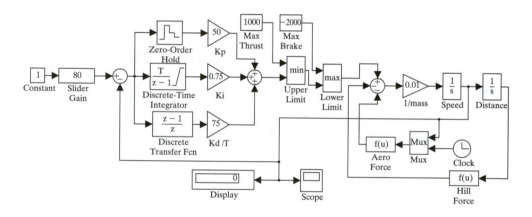

FIGURE 6.12: Car model with discrete controller

6.6 SUMMARY

In this chapter, we discussed using Simulink to model discrete-time systems. We started with scalar linear discrete-time systems, discussing the Unit Delay, Discrete-Time Integrator, and Discrete Transfer Fcn blocks in particular. We briefly discussed vector discrete systems. We concluded with discussions of multirate discrete-time systems and hybrid systems.

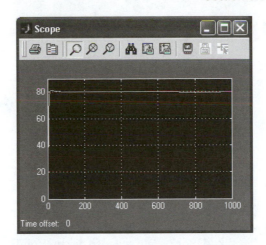

FIGURE 6.13: Car speed controller performance

REFERENCES

[1] Kuo, Benjamin C., *Automatic Control Systems* (Englewood Cliffs, N.J.: Prentice Hall, 1995). This text covers design of continuous and discrete control systems and includes lots of practical examples, including MATLAB scripts. Although Simulink is not used, the block diagrams are easily converted to Simulink models.

[2] Ogata, Katsuhiko, *Designing Linear Control Systems with MATLAB* (Englewood Cliffs, N.J.: Prentice Hall, 1993). This book presents brief tutorials and MATLAB implementations of several important linear systems design techniques, including pole placement, state observers, and linear quadratic regulators.

[3] Ogata, Katsuhiko, *Discrete-Time Control Systems*, 2nd ed. (Englewood Cliffs, N.J.: Prentice Hall, 1994).

[4] Orfanidis, Sophocles J., *Introduction to Signal Processing* (Englewood Cliffs, N.J.: Prentice Hall, 1996). This book provides detailed coverage of Z-transforms, transfer functions, and digital filter design.

[5] Phillips, Charles L., and Nagle, H. Troy, *Digital Control System Analysis and Design*, 3rd ed. (Englewood Cliffs, N.J.: Prentice Hall, 1995). This book provides comprehensive coverage of discrete-time control systems analysis and design. It presents many practical examples and makes good use of MATLAB.

[6] Shahian, Bahram, and Hassul, Michael, *Control System Design Using MATLAB* (Englewood Cliffs, N.J.: Prentice Hall, 1993). This book provides an introduction to MATLAB programming, and uses MATLAB to solve many of the standard problems in classical control, and modern control theory.

[7] Strum, Robert D., and Kirk, Donald E., *Contemporary Linear Systems Using MATLAB* (Pacific Grove, Ca.: Brooks/Cole, 2000).

7

Subsystems and Masking

In this chapter, we will explore some of the ways we can use Simulink to model more complex systems. We will build hierarchical models and develop custom Simulink blocks called masked blocks. We will also discuss conditionally executed subsystems that can make Simulink models more efficient.

7.1 INTRODUCTION

In the preceding chapters, we discussed the basics of building Simulink models for continuous, discrete, and hybrid systems. Using the procedures we covered in the proceeding chapters, it is possible to model any physical system. However, as your Simulink models become more complex, additional Simulink capabilities and programming techniques can make the models easier to develop, to understand, and to maintain. In this chapter, we'll start with a discussion of Simulink subsystems, which provide a capability within Simulink similar to subprograms in traditional programming languages. Next, you'll learn to use masking to make subsystems easier to use and understand. Last, we will discuss conditionally executed subsystems, which facilitate the development of models with multiple modes or phases of operation.

7.2 Simulink SUBSYSTEMS

Most engineering programming languages include the capability to employ subprograms. In FORTRAN, there are subroutine subprograms and function subprograms. C subprograms are called functions; MATLAB subprograms are called function M-files. Simulink provides an analogous capability called *subsystems*. There are two important reasons for using subprograms: abstraction and software reuse.

As models grow larger and more complex, they can easily become difficult to understand and maintain. Subsystems solve this problem by breaking a large model into a hierarchical set of smaller models. As a simple example, consider the automobile model of Example 5.5. The Simulink model is repeated in Figure 7.1. The model consists of two main parts: the automobile dynamics and the controller. When examining the model, it is not clear which blocks represent the automobile dynamics and which blocks constitute the controller. In Figure 7.2, we converted the automobile and controller portions of the model into subsystems. In this version, the conceptual structure is clear in the top level of the model Figure 7.2,

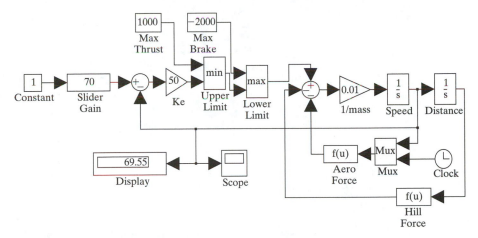

FIGURE 7.1: Automobile model with proportional speed control

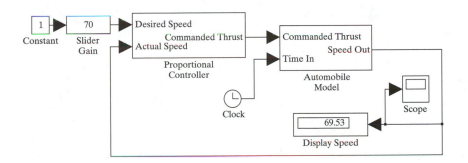

FIGURE 7.2: Hierarchical automobile model

but the details of the controller and automobile dynamics are hidden in the subsystems (Figure 7.3). This hierarchical structure is an example of software abstraction.

Subsystems can also be viewed as reusable model components. Suppose that we wish to compare several different controller designs using the same automobile dynamics model. Rather than building a complete new block diagram each time, it is more convenient to build only the part of the model that is new each time—the controller. Not only does this save time building the model, but it also ensures that we are using exactly the same automobile dynamics. An important advantage of software reuse is that once we verified that a subsystem is correct, we don't have to repeat the testing and debugging process each time we use the subsystem in a new model. Subsystems greatly simplify the task of modeling physical systems that contain several instances of a particular component, such as, for example, the four tire models that would be required to model the ride characteristics of an automobile.

There are two methods to use to build Simulink subsystems. The first method is to encapsulate a portion of an existing model in a subsystem using **Edit:Create**

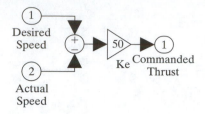

(a) Controller subsystem

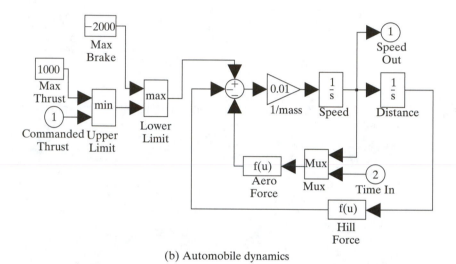

(b) Automobile dynamics

FIGURE 7.3: Hierarchical automobile model

Subsystem. The second method is to use a Subsystem block from the Ports & Subsystems block library. We'll discuss both methods.

7.2.1   Encapsulating a Subsystem

To encapsulate a portion of an existing Simulink model into a subsystem, proceed as follows:

Select all the blocks and signal lines to be included in the subsystem using a bounding box. *Note that you must use a bounding box in this instance.* It is frequently necessary to rearrange some blocks so that you can enclose only the desired blocks in the bounding box.

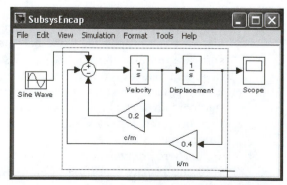

Choose **Edit:Create Subsystem** from the model window menu bar. Simulink will replace the selected blocks with a Subsystem block with an input port for each signal entering the new subsystem and an output port for each signal leaving the new subsystem.

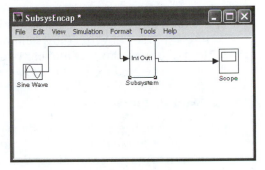

Simulink will assign default names to the input and output ports.

Resize the Subsystem block so that the port labels are readable, and rearrange the model as desired.

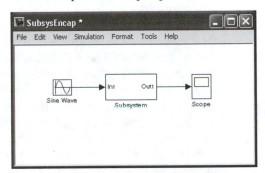

To view or edit the subsystem, double-click the block. A new window will appear that contains the subsystem. In addition to the original blocks, an Inport block is added for the signal entering the subsystem and an Outport block is added for the signal exiting the subsystem. Changing the labels on these ports changes the labels on the new block's icon. Click the control to close the subsystem window when you've finished editing the subsystem.

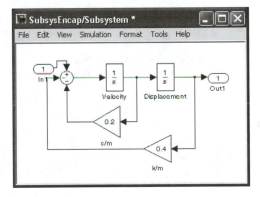

Edit:Create Subsystem does not have an inverse operation. Once you encapsulate a group of blocks into a subsystem, there is no menu choice to reverse the process. Therefore, it is a good idea to save the model before creating the subsystem. If you decide you don't want to accept the newly created subsystem, close the model window without saving, then reopen the model. To manually reverse the encapsulation of a subsystem, copy the subsystem to a new model window, open the subsystem, then copy the blocks from the subsystem window to the original model window.

By default, subsystems are *virtual*. When the Simulink model is executed, Simulink evaluates blocks according to a default evaluation order, just as if all the blocks in the subsystem were in the base model window. In some models, it may be desirable, in terms of execution order, to treat a subsystem as a single block, and therefore to evaluate all blocks in the subsystem as a group called an *atomic subsystem*. To convert a subsystem into an atomic subsystem, select the subsystem block and then choose **Edit:SubSystem Parameters** from the model window menu bar. Select checkbox **Treat as atomic unit**.

7.2.2 Subsystem Blocks

If, when building a model, you know that you will need a subsystem, you may find it convenient to build the subsystem in a subsystem window directly. This eliminates the need to rearrange the blocks that will compose the subsystem to fit in a bounding box. It also avoids having to tidy up the model window after the subsystem is encapsulated.

To create a new subsystem using a Subsystem block, drag a Subsystem block from the Ports & Subsystems block library to the model window. Double-click the Subsystem block. The subsystem window will appear. Build the subsystem using the standard procedures for constructing a model. Use Inport blocks for all signals entering the subsystem and Outport blocks for all signals leaving the subsystem. If desired, change the labels on the Inport and Outport blocks to identify the purpose of each input and output. Close the subsystem window when you've finished building the subsystem. Note that you do not need to choose **File:Save** before closing the subsystem window; the subsystem is part of the model in which the subsystem is created, and it is saved when that model is saved.

EXAMPLE 7.1 Spring-mass subsystem

We wish to model the spring-mass system composed of carts connected as shown in Figure 7.4. We will build the model from subsystem blocks that model each cart as shown in Figure 7.5.

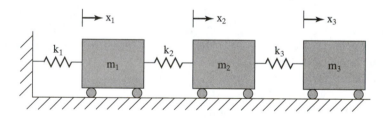

FIGURE 7.4: Spring-mass system

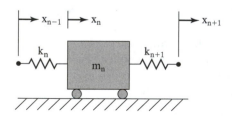

FIGURE 7.5: Single-cart model

The equation of motion for a single cart is

$$\ddot{x}_n = \frac{1}{m_n}\left[k_n\left(x_{n-1} - x_n\right) - k_{n+1}\left(x_n - x_{n+1}\right)\right]$$

Using the procedure discussed in Section 7.2.2, construct the subsystem as shown in Figure 7.6. This subsystem will model Cart 1. The inputs to the single cart subsystem are x_{n-1} (position of the cart to the left) and x_{n+1} (position of the cart to the right). The subsystem output is x_n (position of cart). Notice that each spring is referenced in two subsystem blocks: one for the cart to the right of the spring and one for the cart to the left.

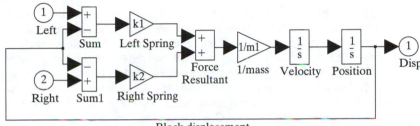

FIGURE 7.6: Cart model subsystem for Cart 1

Once the subsystem is complete, close the subsystem window. Make two copies of the subsystem block, and connect the blocks as shown in Figure 7.7.

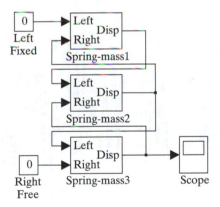

FIGURE 7.7: Three-cart model using subsystems

It is convenient in this case to enter the spring constants (k1, k2, k3) and cart masses (m1, m2, m3) as MATLAB variables, and assign values to the variables using a MATLAB script M-file (which we named SetCartParms.m), as shown in Listing 7.1. Execute this script M-file from the MATLAB prompt before running the simulation. (Note that the script M-file (extension .m) must have a name that is different from the name of the Simulink model. For example, if the Simulink model is named examp_1.mdl, and you name the script M-file examp_1.m, MATLAB will open the Simulink model when you enter the command examp_1 at the MATLAB prompt.)

Listing 7.1: MATLAB script `SetCartParms.m` to initialize spring constants

```
% Set the spring constants and block mass values
k1 = 1 ;
k2 = 2 ;
k3 = 4 ;
m1 = 1 ;
m2 = 3 ;
m3 = 2 ;
```

The block parameters for each block in each copy of the subsystem must now be set. For Cart 1, set the value of **Gain** for the Gain block labeled `Left Spring` to k1, and for Gain block `Right Spring` to k2. Next, set **Gain** for the Gain block labeled `1/mass` to `1/m1`. Initialize the `Velocity` Integrator block to 0, and the `Position` Integrator block to 1.

For Cart 2, set the value of **Gain** for the Gain block labeled `Left Spring` to k2, and for Gain block `Right Spring` to k3. Next, set **Gain** for the Gain block labeled `1/mass` to `1/m2`. Initialize the `Velocity` Integrator block to 0 and the `Position` Integrator block to 0.

For Cart 3, set the value of **Gain** for the Gain block labeled `Left Spring` to k3, and for Gain block `Right Spring` to 0, because there is no right spring for this cart. Next, set **Gain** for the Gain block labeled `1/mass` to `1/m3`. Initialize the `Velocity` Integrator block to 0, and the `Position` Integrator block to 0.

We configured the Scope block to save the scope data to the workspace, and set the simulation **Start** time to 0 and **Stop** time to 100. After running the simulation, the scope data were plotted from within MATLAB, resulting in the trajectory plot for Cart 3 shown in Figure 7.8.

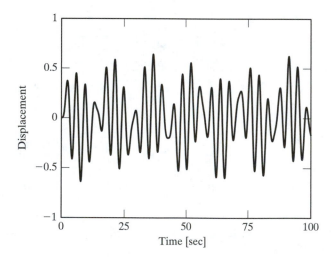

FIGURE 7.8: Trajectory of Cart 3

7.3 MASKED BLOCKS

Masking is a Simulink capability that extends the concept of abstraction. Masking permits us to treat a subsystem as if it were a simple block. A masked block may have a custom icon, and it may also have a dialog box in which configuration parameters are entered in the same way parameters are entered for blocks in the Simulink block libraries. The configuration parameters may be used directly to initialize the blocks in the underlying subsystem, or they may be used to compute data to initialize the blocks.

To understand the concept of masking, consider the model shown in Figure 7.9(a). This model is equivalent to the model in Example 7.1, but it is easier to use. Double-clicking the block labeled Spring–mass 1 opens the dialog box shown in Figure 7.9(b). Instead of opening the dialog box for each Gain block and each Integrator to set the block parameters, you can enter all the parameters for each subsystem in the subsystem's dialog box. The dialog box in Figure 7.9(b) "masks" a subsystem that is nearly identical to the subsystem in Figure 7.6.

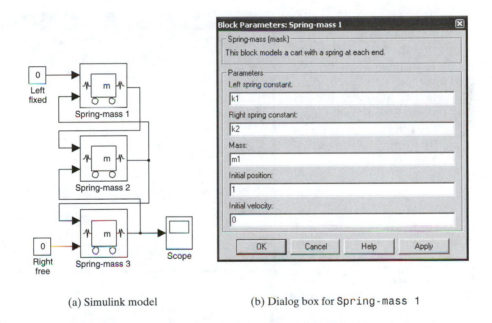

(a) Simulink model   (b) Dialog box for Spring-mass 1

FIGURE 7.9: Three-cart model using masked subsystems

In this section, we will explain the steps in creating a masked subsystem. The examples will show how to create the spring-mass masked subsystem. Additional examples will illustrate other masking features.

The process of producing a masked block can be summarized as follows:

1. Build a subsystem using the procedures discussed in Section 7.2.

2. Select the subsystem block, then choose **Edit:Mask Subsystem** from the model window menu bar.

3. Using the Mask Editor, set up the mask documentation, dialog box, and optionally, build a custom icon.

7.3.1   Converting a Subsystem into a Masked Subsystem

The first step in creating a masked subsystem is to create a subsystem using the procedures described in Section 7.2. To illustrate the process, let's build a spring-mass masked block starting with one of the subsystems in the model in Figure 7.7.

Open the model shown in Figure 7.7. Next, open a new model window. Drag a copy of the block labeled `Spring-mass` 1 to the new model window. Select the block, then choose **Edit:Mask Subsystem** from the new model window menu bar.

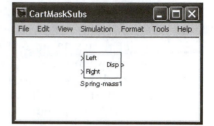

The Mask Editor dialog box will appear. Note that the Mask Editor has four tabbed pages. We will discuss each page in the following subsections. Before proceeding, save the new model window using the name Cart-MaskSubs.

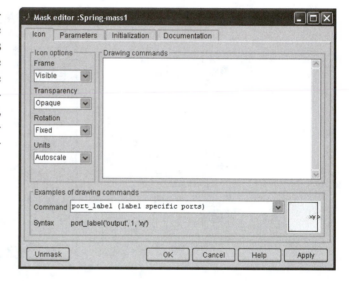

7.3.2   Mask Editor Documentation Page

The Documentation page is illustrated in Figure 7.10(a). The page consists of three fields, which in Figure 7.10(a) were filled in for the spring-mass block. All fields in the Documentation page are optional.

After filling in the fields as shown, select **Close**. Then, double-click the spring-mass block, opening the dialog box shown in Figure 7.10(b). You can return to the Mask Editor by selecting the block, then choosing **Edit:Edit Mask** from the model window menu bar.

We will discuss each field in the Documentation page next.

Mask Type Field.   The contents of the first field, **Mask type**, will be displayed as the block type in the masked block's dialog box. Notice that there are two labels in the upper-left corner of the block dialog box [Figure 7.10(b)]. The label in the window title bar (here `Spring-mass1`) is the label of the currently selected block. The label inside the dialog box (here `Spring-mass`) is the block type. Every instance of this new block will have the same block type, but each instance

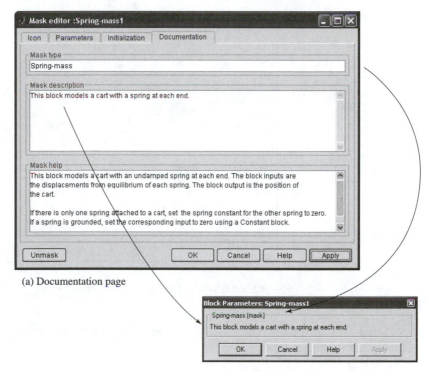

(a) Documentation page

(b) Corresponding masked block dialog box

FIGURE 7.10: Mask Editor **Documentation** page

in a particular model must have a different label. Also, note the word "mask" in parentheses appended to block type, indicating that this is a masked block. (Compare the masked block dialog box to the dialog box for a block from a block library.)

Block description field.   The second field, **Block description**, is displayed in a bordered area at the top of the masked block's dialog box. This area should contain a brief description of the block's purpose and any needed reminders concerning the use of the block.

Block help.   The third field, **Block help**, will be displayed by the MATLAB Help system when the masked block's dialog box **Help** button is pressed. This field should contain detailed information concerning the use, configuration, and limitations of the masked block and the underlying subsystem.

7.3.3 Parameters Page

The **Parameters** page (Figure 7.11) is used to define parameters for blocks in the subsystem underlying the masked block. The **Parameters** page can be divided into two sections. The top section defines masked block dialog fields and the order in which they are displayed in the masked block dialog box and associates a MATLAB

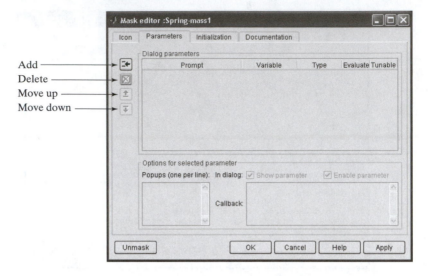

FIGURE 7.11: Mask Editor **Parameters** page

variable with each field. The bottom section contains certain options for each field defined in the top section.

Dialog Parameters.   The top section of the Mask Editor **Parameters** page is used to create, edit, and delete dialog box fields and associate a MATLAB variable with each field. This section consists of a scrolling list of dialog box fields, buttons to add, delete, and move fields, and five columns used to configure the masked block dialog box fields and associated MATLAB variables. To add the first prompt field in the Spring-mass block dialog box, proceed as follows:

Select the block, then open the Mask Editor by choosing **Edit:Edit Mask** from the model window menu bar. Select the **Parameters** page.

Click the **Add** button.

A blank line will be added to the parameter list.

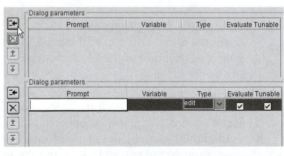

In the **Prompt** field, enter Left spring constant. In the **Variable** field, enter k_left. Notice that the prompt and the variable name are displayed in the parameter list.

Choose **OK**. Double-click the Spring-mass subsystem. The block dialog box now has a prompt. The value entered in the field corresponding to the prompt will be assigned to MATLAB variable k_left.

Add prompt Right spring constant with variable k_right, prompt Mass with variable mass, prompt Initial position with variable x0, and prompt Initial velocity with variable x_dot0.

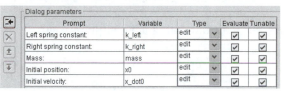

That completes the process of creating the dialog box fields. To see the results, press **OK** to save the changes and exit the Mask Editor.

Double-click the Spring-mass block, opening the block dialog box, which now includes all five prompts.

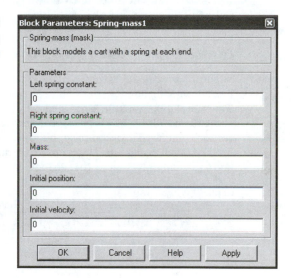

Now, we will discuss in more detail the buttons and fields in the dialog box prompt section of the Mask Editor. First, notice that there are four buttons on the left side of the dialog box prompt section. These buttons are used to add, delete, and arrange dialog box prompts. To create a new field, click the line in the scrolling parameter list of the item you wish to precede the new field. Then, click the **Add** button. A blank line will appear in the scrolling list. Clicking the **Delete** button deletes the selected field. The **Up** and **Down** buttons move the selected field in the appropriate direction in the list of fields. So, if you wish for the new field to be the top field, click on first parameter, add a line, and then move the new line to the top of the list.

The third column of the parameter list, **Type**, is a drop-down list with three options: Edit, Checkbox, and Popup. Edit produces a field in which data is entered (a fill-in-the-blank field), and it is the most common type of field. Checkbox generates a field that has two possible values, depending on whether or not the box is checked. Popup produces a list of choices set in **Popup** in the Options section.

The value assigned to the internal variable associated with a block dialog box field will depend upon the state of the corresponding check box in the **Evaluate** column. If **Evaluate** is checked, the variable associated with the field will contain the value of the expression in the field. So, for example, if the field contains k1, and in the MATLAB workspace k1 is assigned the value 2.0, the variable associated with the dialog field will be assigned the value 2.0. If **Evaluate** is not checked, the variable associated with the dialog field will contain the character string 'k1'.

The last column, **Tunable**, determines whether the parameter may be changed during the execution of a simulation. If **Tunable** is checked, the parameter may be changed during a simulation, otherwise, the parameter stays constant during a simulation.

Setting **Type** to Checkbox produces a check-box field. The variable associated with a check-box field will be assigned a value depending on the setting of **Evaluate**. If **Evaluate** is selected, the variable associated with the field will be set to 0 if not checked or 1 if checked. If **Evaluate** is not selected, the variable associated with the field will be set to 'off' if not checked and 'on' if checked.

EXAMPLE 7.2   Using a check box

Suppose we wish to configure a masked subsystem such that its block dialog box has a check box allowing the user to specify that angular inputs are to be in degrees rather than radians. The value of the variable associated with the check box (c_stat) is to be 0 if the box is not checked and 1 if the box is checked. Shown in Figure 7.12(a) is the Mask Editor Parameters section configured to accomplish this task, and shown in Figure 7.12(b) is the corresponding masked block dialog box. Here, on the **Documentation** page, **Mask type** is set to Get units and Mask description to This block illustrates a check box.

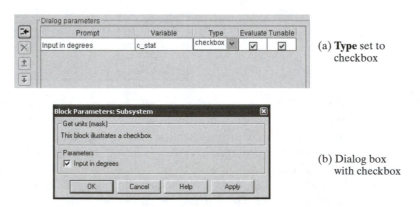

(a) **Type** set to checkbox

(b) Dialog box with checkbox

FIGURE 7.12: Mask Editor **Documentation** page

If **Type** is set to Popup, the field **Popups** in the options section is used to define a list of choices, as will be shown in Example 7.3. The variable associated with a popup field will be assigned a value depending upon the setting of the corresponding **Popups** check box. If **Popups** is checked, the variable associated with the field will be set to the ordinal number of the selected popup choice. So, for example, if the first choice is selected, the value of the variable will be set to 1. If the second choice is selected, the value of the variable will be set to 2, and so on. If **Popups** is not checked, the variable will contain the character string corresponding to the selected choice. Enter the popup choices in field **Popups**, one to a line.

EXAMPLE 7.3   Using a popup control

Suppose we wish to create a masked block that produces an output signal defined by a popup list containing choices Very hot, Hot, Warm, Cool, and Cold. To do this, open a new model, then drag a Constant block into the model window. Select the Constant block using a bounding box, then choose **Edit:Create Subsystem** from the model window menu bar. Select the new subsystem, then choose **Edit:Mask Subsystem** from the model window menu bar. Set **Mask type** to Popup example, and **Block description** to This block illustrates a popup list. Configure the **Parameters** page as shown in Figure 7.13(a). Choose **OK**. Double-click the masked block, then click on the popup control. The block dialog box should be as shown in Figure 7.13(b).

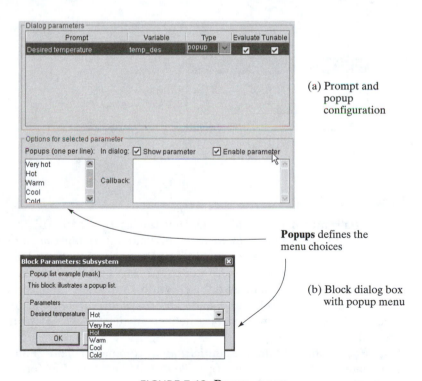

(a) Prompt and popup configuration

Popups defines the menu choices

(b) Block dialog box with popup menu

FIGURE 7.13: Popup menu

Callback Field.   The **Callback** field allows you to associate with the selected parameter a block of MATLAB code called a callback that is executed when the dialog parameter is entered. A detailed discussion of callbacks is presented in Chapter 9. Although callbacks can be used for many purposes in Simulink models, the masked block parameter callback is most useful for performing error and consistency checking on parameters. For example, the callback illustrated in Figure 7.14(a) will test the value entered for the left spring constant for the spring-mass block (k_left) and produce the error dialog in Figure 7.14(b) if a negative value is entered.

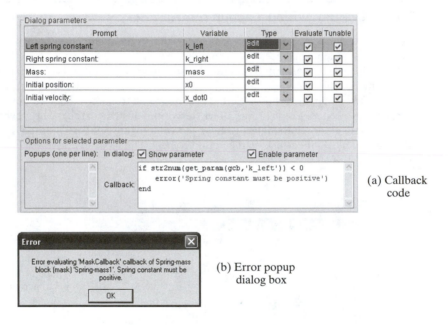

(a) Callback code

(b) Error popup dialog box

FIGURE 7.14: Parameter callback code

7.3.4 Initialization Page

The **Initialization** page provides a list of variables associated with the block parameters and field **Initialization commands**. This field can contain one or more MATLAB statements that assign values to MATLAB variables that can be used to configure blocks in the masked subsystem. The MATLAB statements may use any MATLAB operator, built-in or user-written functions, and control flow statements such as `if`, `while`, and `end`. The scope of variables in the **Initialization commands** field is local; variables defined in the MATLAB workspace are not accessible.

Each command in the **Initialization commands** field should normally be terminated with a semicolon (;). If you omit the semicolon for a command, the results of the command will be displayed in the MATLAB window whenever the command is executed. This provides a convenient means for debugging the commands.

EXAMPLE 7.4   Configuring Initialization commands

The Spring-mass block icon will need two wheels. To prepare to draw the wheels, create two vectors: one containing the x coordinates of a small circle and the other the y coordinates. These variables will be used in the Icon page to draw the wheels.

```
Initialization commands
t=[0:0.5:2*pi,0];
x=cos(t);
y=sin(t);
```

Configuring Subsystem Blocks.   The blocks in the masked subsystem must be configured to use the variables defined on the **Initialization** and **Parameters** pages. To configure the blocks in the Spring-mass subsystem, select the subsystem, then choose **Edit:Look Under Mask** from the model window menu bar. Double-click the Gain block labeled `Left spring`, and set **Gain** to `k_left`. Likewise, set the value of **Gain** for Gain block `Right spring` to `k_right` and for Gain block `1/mass` to `1/mass`. Set **Initial condition** for Integrator `Velocity` to `x_dot0` and Integrator `Position` to `x0`. The subsystem should now appear as shown in Figure 7.15. Close the subsystem, and save the model.

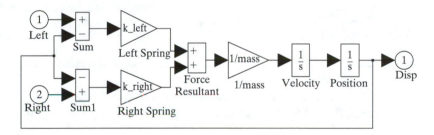

FIGURE 7.15: Spring-mass subsystem configured to use masked block variables

Local Variables.   An important difference between masked subsystems and nonmasked subsystems is the scope of variables in the dialog boxes for the blocks in a subsystem. Blocks in nonmasked subsystems may use any MATLAB variable currently defined in the MATLAB workspace. This feature was used to initialize the subsystems in Example 7.1. The blocks in a masked subsystem cannot access variables in the MATLAB workspace; a masked subsystem has its own internal name space that is independent of the MATLAB workspace and all other masked subsystems in a Simulink model. This is an extremely valuable feature of masked subsystems, as it eliminates the possibility of unintentional variable name conflicts.

A masked subsystem's internal variables are created and assigned values using the Mask Editor dialog fields and initialization commands. Each dialog field in a masked block's dialog box defines an internal variable accessible only within the masked subsystem. Additional internal variables may be defined in the **Initialization commands** field of the **Initialization** page.

The connections between the MATLAB workspace and a masked subsystem are the contents of the masked block's dialog box fields. An input field in a masked block's dialog box may contain constants or expressions using variables defined in the MATLAB workspace. The value of the contents of the input field is assigned to the masked subsystem internal variable associated with the input field. This internal variable may be used to initialize a block in the masked subsystem, or it may be used to define another internal variable defined in the **Initialization commands** field.

Consider the model shown in Figure 7.9, with the masked subsystem configured as shown in Figure 7.15. The spring constant for the left spring in each instance of the masked subsystem is k_left. However, the contents of k_left in each instance is different. k_left for the first Spring-mass block should be set (using the Spring-mass block dialog box) to k1. k_left for the second block should be set to k2 and for the third Spring-mass block to k3.

EXAMPLE 7.5   Using internal variables in a masked subsystem

To illustrate the use of internal variables in a masked subsystem, consider the assignment of a value to the spring constant of the right spring in the cart subsystem of Figure 7.9. Earlier, the contents of the cart subsystem dialog field **Right spring constant** were associated with internal variable k_right. In Figure 7.16, we see that the Gain block labeled Right spring is set to k_right. Thus, when the M-file script SetCartParms.m is executed, the value of the **Gain** for the Gain block Right spring in this particular instance of the cart subsystem is set to 2.

EXAMPLE 7.6   Setting an internal variable using a popup list

For the masked subsystem in Example 7.3, define a variable temp_val as follows:

Menu Choice	temp_val
Very hot	120
Hot	100
Warm	85
Cool	70
Cold	50

To accomplish this task, place the following statements in the **Initialization commands** field:

```
temp_list = [120,100,85,70,50] ;
temp_val = temp_list(temp_des) ;
```

The first statement creates a vector of temperatures. The second statement uses temp_des [the variable associated with the pop-up list in Example 7.3 (Figure 7.13)] as an index into the vector. Choose **Close**, and then, with the masked block still selected, choose **Edit:Look Under Mask**, and set Constant block dialog box field **Constant value** to temp_val.

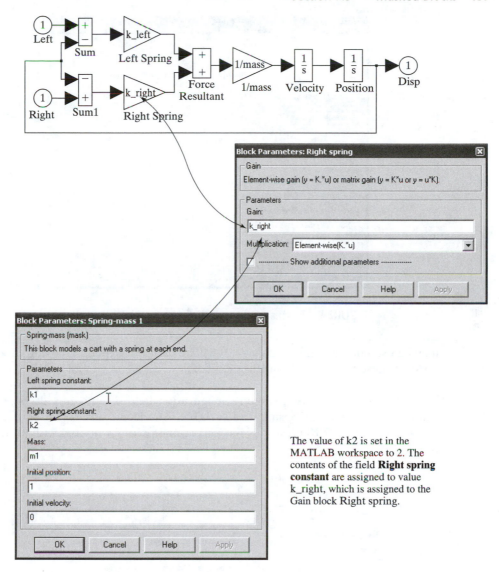

FIGURE 7.16: Right spring constant Gain block initialization

7.3.5   Mask Editor Icon Page

The **Icon** page allows you to design custom icons for masked blocks. The **Icon** page used to create the custom icon for the cart block in Figure 7.7 is shown in Figure 7.17. (Recall that x and y were defined in the **Initialization commands** field in Example 7.4.) The page consists of six fields. **Drawing commands** is a multiple line field in which you may enter one or more MATLAB statements to draw and label the icon. The four fields to the left of **Drawing commands** configure the block icon. The final field, **Command**, is a drop-down list of available MATLAB commands that

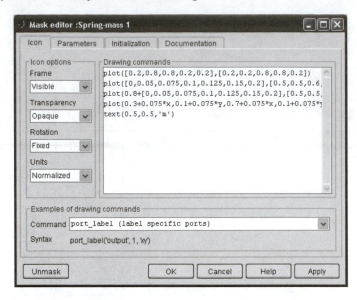

FIGURE 7.17: Mask Editor Icon page for the cart subsystem

may be used in **Drawing commands**. It will be easier to explain **Drawing commands** if we first discuss the icon configuration fields.

Frame Field. The first icon configuration field, **Frame**, is a drop-down list containing two choices: **Visible** and **Invisible**. The icon frame is the border of the block icon. Figure 7.18 illustrates the Spring-mass block with and without the icon frame.

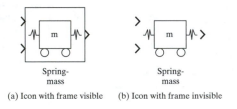

(a) Icon with frame visible (b) Icon with frame invisible

FIGURE 7.18: Icon frame visibility

Transparency Field. The second icon configuration field, **Transparency**, is a drop-down list containing two choices: **Transparent** and **Opaque**. Shown in Figure 7.19 is the Spring-mass block with **Transparency** set to both options. Note that when **Transparent** is selected, the labels on the Inport and Outport blocks in the subsystem underlying the mask are visible. Selecting **Opaque** hides the labels.

Rotation Field. The third icon configuration field, **Rotation**, is a drop-down list containing two choices: **Fixed** and **Rotates**. This field determines the behavior of the block icon when **Format:Flip block** and **Format:Rotate block** are selected.

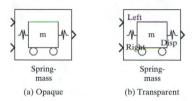

FIGURE 7.19: Icon transparency

If **Fixed** is selected, when the block is rotated or flipped the icon orientation doesn't change. When **Rotates** is selected, the orientation of the icon is the same as the orientation of the block. Illustrated in Figure 7.20 is the difference. **Fixed** is frequently desirable, particularly in cases in which the block icon contains text.

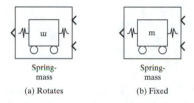

FIGURE 7.20: Icon rotation

Units Field.   The final icon configuration field, **Units**, determines the scale used in plotting icon graphics and locating text on the icon. The field is a drop-down list consisting of three choices: **Autoscale**, **Pixels**, and **Normalized**.

Pixels is an absolute scale and will result in an icon that isn't resized when the block is resized. The coordinates of the lower-left corner of the icon are (0,0). The units are pixels, and therefore, the size of the icon will depend on the display resolution. Figure 7.21 shows a masked block with **Units** set to **Pixels**.

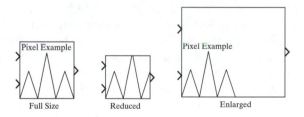

FIGURE 7.21: Units set to Pixels

Autoscale adjusts the size of the icon to exactly fit in the block's frame (even if the frame is invisible). Figure 7.22 shows the masked block in Figure 7.21 reset to **Autoscale**. Note that the text on the icon does not change size when the block is resized.

Normalized specifies that the drawing scale is 0.0 to 1.0 in both the horizontal and vertical axes. The coordinates of the lower left corner of the icon (in its default

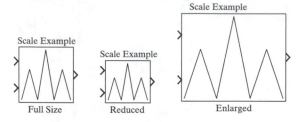

FIGURE 7.22: Units set to Autoscale

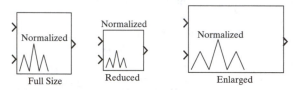

FIGURE 7.23: Units set to Normalized

orientation, not rotated or flipped) are defined to be (0,0), and the coordinates of the upper right corner of the icon are (1,1). When the block is resized, the coordinates are also resized. Text does not change size when the block is resized. Figure 7.23 illustrates a masked block with **Units** set to **Normalized**.

Drawing Commands. Several different MATLAB statements may be entered in Drawing commands to customize a block's icon. Table 7.1 lists these commands.

Three of the commands display text on the icon. The simplest, `disp(string)`, displays `disp(string)` centered on the icon. This command is useful for placing a simple descriptive title in the center of the icon. The `text(x,y,string)` command permits you to locate a string anywhere on the icon, using the icon coordinate system as specified in the Drawing coordinates field. The third text command, `fprintf(string,list)` is identical to the MATLAB `fprintf` statement. (Enter `help fprintf` at the MATLAB prompt for details on `fprintf`.) Using `fprintf`, you can build labels that use variables defined by the block's dialog fields and the statements in the field **Initialization commands** on the **Initialization** page. Embed a new-line character (\n) in the string to produce a label with multiple lines. Like the `disp` command, `fprintf` places the label at the center of the icon.

A character string used to display text on the icon may be a literal string, or it may be a MATLAB string variable. A literal string is a sequence of printable characters enclosed in single quotes. For example, to place the label "Special Block" in the center of a block's icon, use the command:

```
disp('Special Block')
```

A string variable is a MATLAB variable that represents a character string instead of a number. MATLAB provides several functions for building and manipulating string variables. These functions can be used in the field **Initialization commands** on the **Initialization** page to create strings for use in the text display

TABLE 7.1: Icon Drawing Commands

Command	Description
disp(*string*)	Display *string* in the center of the icon.
text(*x,y,string*)	Display *string* starting at (*x,y*).
text(*x,y,string*, 'HorizontalAlignment', *halign*, 'VerticalAlignment', *valign*)	Display *string* relative to (*x,y*). The two alignment options allow you to control the positioning of the text relative to the location (*x,y*). Either or both option pairs may be used. If option 'HorizontalAlignment' is specified, parameter *halign* must immediately follow it and must be a string or string variable containing 'left', 'right', or 'center'. If option 'VerticalAlignment' is specified, parameter *valign* must immediately follow it and must be a string or string variable containing 'base', 'bottom', 'middle', 'cap', or 'top'.
fprintf(*string,list*)	Display the results of the fprintf statement at the center of the icon.
plot(*x_vector,y_vector*)	Draw a plot on the icon.
dpoly(*num,denom*)	Display a transfer function centered on the icon.
dpoly(*num,denom*'z')	Display a discrete transfer function in ascending powers of *z*.
dpoly(*num,denom*'z-')	Display a discrete transfer function in descending powers of *z*.
droots(*zeros,poles, gain*)	Display a transfer function in zero-pole-gain format.
port_label(*type, number, label*)	Display a label on a port. *type* must be 'input' or 'output'. *number* is the input or output port number. *label* is a text string or string variable containing the desired port label. Note that if there are any icon drawing commands, the port labels on the subsystem input and output ports are not displayed. A separate port_label command must be used for each port for which a label is desired.
image(*object*, [*x, y, width, height*])	Display image at position (*x, y*) relative to the lower-left corner of the icon.
image(*object*, [*x, y, width, height*])	Display image at position (*x, y*) relative to the lower-left corner of the icon.
image(*object*, [*x, y, width, height*], *rotation*)	The additional parameter *rotation* must be a string or string variable containing either 'on' or 'off'. If 'on', the image rotates if the block is rotated. If 'off' (the default), the image does not rotate.

TABLE 7.1: Icon Drawing Commands (Cont)

Command	Description
patch (*x*, *y*)	Draws a closed polygon with vertices defined by the *x* and *y* vectors. The *x* and *y* vectors must be of the same length and must have at least three elements. The polygon will be drawn in the current foreground color.
patch (*x*, *y*, [*red green blue*])	Overrides the current foreground color with specified values of red, green, blue.
color(*color*)	Set the drawing color for subsequent commands to the specified color.

commands. A particularly useful string function is sprintf. sprintf is similar to fprintf, but it writes to a character string instead of the screen or a file. An excellent reference for a detailed discussion of MATLAB string variables and functions is the text by Hanselman and Littlefield [1].

EXAMPLE 7.7   Displaying a parameter value on a block icon

Suppose we wish to change the icon shown in Figure 7.9 such that the cart mass is displayed on the icon just above the wheels, and the units of mass are kilograms. Add the following command to the **Initialization commands** field on the **Initialization page** of the Mask Editor:

```
b_label=sprintf('%1.1f kg',mass);
```

Then, enter the following command in the **Drawing commands** field of the **Icon** page:

```
ext(0.25,0.4,b_label);
```

The block icon will be as shown in Figure 7.24.

FIGURE 7.24: Displaying the cart mass on its icon

The plot(*x_vector*,*y_vector*) command displays graphics on the block icon. This command is similar to the MATLAB plot command but has fewer options. In particular, the Mask Editor plot command does not support options to set line styles or colors and will not plot two-dimensional arrays. The command expects pairs of vectors specifying sequences of *x* and *y* coordinates. There may be

more than one pair of vectors in a single plot command, and there may be more than one plot command for an icon. Figure 7.17 shows the commands used to create the cart icon.

EXAMPLE 7.8   **Drawing on a block icon**

Suppose we wish to display sine and cosine functions on an icon. Placing the following commands in the **Initialization commands** field will produce the necessary vectors:

```
x_vector = [0:0.05:1];
y_sin = 0.5 + 0.5*sin(2*pi*x_vector) ;
y_cos = 0.5 + 0.5*cos(2*pi*x_vector) ;
```

Next, place the following command in **Drawing commands**:

```
plot(x_vector,y_sin,x_vector,y_cos) ;
```

The block will appear as shown in Figure 7.25.

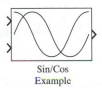

Sin/Cos
Example

FIGURE 7.25: Drawing curves on a block icon

EXAMPLE 7.9   **Drawing logic gate icons**

The Simulink Logical Operator block can be configured to implement AND and OR gates, but the block icon is a rectangle with the logical operator it implements displayed on the icon. By placing copies of the Logical Operator block into masked subsystems, we can produce AND and OR blocks that more closely resemble the conventional icons for these gates.

To produce the AND gate block, start with a Logical Operator block configured to implement AND. Select the block using a bounding box, then choose **Edit:Create subsystem** from the model window menu bar. Select the subsystem, then choose **Edit:Mask Subsystem** from the model window menu bar. Open the Mask Editor, and place the following command in the **Initialization commands** field on the **Initialization** page:

```
t=-pi/2:0.1:pi/2;
```

Change to the **Icon** page, and set **Frame** to **Invisible** and **Units** to **Normalized**. Place the following commands in **Drawing commands**:

```
plot([0.5,0,0,0.5],[0,0,1,1],0.5+0.5*cos(t),0.5+0.5*sin(t));
port_label('input', 1, 'a');
port_label('input', 2, 'b');
port_label('output', 1, 'ab');
```

To produce the OR gate block, the process is similar, using the following in **Drawing commands**:

```
plot([0,0],[0,1],t,0.5*t.^2,t,1-0.5*t.^2);
port_label('input', 1, 'a');
port_label('input', 2, 'b');
port_label('output', 1, 'a+b');
```

The logic gate blocks that result are shown in Figure 7.26.

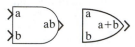

FIGURE 7.26: Logic gate masked blocks

The drawing commands image(*object*), image(*object*, [*x, y, width, height*]), and image(*object*, [*x, y, width, height*], *rotation*) allow you to place color images on block icons. *object* is a MATLAB image usually read using the MATLAB command imread(*ImageFile*), where *ImageFile* is a file containing an image in a format compatible with MATLAB.

EXAMPLE 7.10   Displaying a .tif image on a block icon

Suppose we wish to display a photograph of an aircraft in tagged image format, phantom.tif, on a block icon for a masked subsystem that models the aircraft flight dynamics. Assuming the file phantom.tif is in the current MATLAB path, we can place the following statement in **Drawing commands**:

```
image(imread('phantom.tif'));
```

The icon will appear as shown in Figure 7.27. Note that the image is not stored with the Simulink model—it must be loaded from the file each time the model is opened.

FIGURE 7.27: Placing an image on a block icon

The drawing commands `dpoly(num,denom)` and `droots(zeros,poles, gain)`, display a transfer function on the blocks icon.

`dpoly` displays the transfer function in polynomial form. The arguments *num* and *denom* are vectors containing the coefficients of the transfer function numerator and denominator in descending powers of s. `dpoly` will also display discrete transfer functions in descending powers of z or ascending powers of $1/z$. To display the transfer function in descending powers of z, use the command `dpoly(num,denom,'z')`. To display the transfer function in ascending powers of $1/z$, use `dpoly(num,denom,'z-')`.

`droots` displays a transfer function in factored pole-zero form. *zeros* is a vector containing the zeros of the transfer function (roots of the numerator), and *poles* is a vector containing the poles of the transfer function (roots or the denominator). *gain* is a scalar.

7.3.6   Looking Under and Removing Masks

There are two additional masking commands that permit you to view a subsystem underlying a mask and to delete the mask.

To examine the subsystem underlying a masked block, select the masked block, then choose **Edit:Look under mask**.

To convert a masked block into an unmasked block or subsystem, select the block, then open the Mask Editor. Press the **Unmask** button at the bottom of the Mask Editor. If you change your mind about removing the mask, select the block and choose **Edit:Mask Subsystem**. The masking information will be preserved until you close the model. Once you close the model after removing the mask, it is not possible to restore the mask.

7.3.7   Using Masked Blocks

Once you have created a masked block, it can be copied to a model window in a manner identical to that used to copy a block from a Simulink block library. For example, to build the model shown in Figure 7.9(a), open a new model window, then drag three copies of the Spring-mass block to the new model window. Add the Constant and Scope blocks, and connect the blocks with signal lines as shown. Configure the blocks as shown in Table 7.2

Now the model is complete. Configure the Scope block and simulation parameters as in Example 7.1, then save the model. In the MATLAB workspace, run the

TABLE 7.2: Cart Model Configuration Parameters

Field	Spring-mass 1	Spring-mass 2	Spring-mass 3
Left spring constant	k1	k2	k3
Right spring constant	k2	k3	0
Mass	m1	m2	m3
Initial position	1	0	0
Initial velocity	0	0	0

script M-file (set_x4a.m) to assign values to the spring constants and masses. This model will produce results identical to those of the model in Example 7.1.

7.3.8 Creating a Block Library

A *block library* is a special Simulink model that serves the same purpose as a subroutine library in a conventional programming language. When a block is copied from a block library to a model, the copy remains linked to the version of the block in the block library. If the version of the block in the block library is changed, the change is effective each place the block is used. A block in a block library is called a *library block*. A copy of library block in a model is called a *reference block*. Each reference block has its own data (the dialog box fields), but the functionality is defined by the library block.

A reference block cannot be changed. So, for example, if you select a reference block for which the library block is a masked subsystem, then choose **Edit** from the model window menu bar (the **Edit:Edit Mask** menu choice will not be present). **Edit:Look Under Mask** will be present. But if you look under the mask, and then try to edit the underlying subsystem, Simulink will display an error message.

Creating a Block Library. Create a block library by selecting **File:New:Library** from any model window menu bar or the Simulink block library menu bar. (Recall that you can open the Simulink block library by right-clicking the Simulink Library icon in the Simulink Library Browser window and then clicking the single menu choice Open the 'Simulink' Library.) Copy the desired blocks to the new library, and then save the block library. Once the library has been saved, the blocks in the library are library blocks, and blocks copied from the library are reference blocks.

You can create nested block libraries by adding subsystems to a block library window. For example, each of the block libraries (Sources, Sinks, etc.) in the Simulink block library is a subsystem that contains a set of unconnected blocks. To create a subsystem block library, drag an empty Subsystem block to the block library window. Open the Subsystem block, add the desired blocks to the subsystem window, and delete the default inport and outport blocks and signal lines. A subsystem library can be masked, as are the block libraries in the Simulink block library. The masked subsystem can have a custom icon, but it must not have any dialog box fields (the **Parameters** page should be empty), and the **Block description** field on the **Documentation** page must be blank. Otherwise, double-clicking the block library icon would open a block dialog box, instead of opening a subsystem window from which to copy blocks.

The final step in creating a block library is to add the directory in which the library is stored to the MATLAB path. Reference blocks can find only library blocks that are in the MATLAB path or the current directory (the current MATLAB directory).

Modifying a Block Library. Once a block library has been saved and closed, it is locked and cannot be changed. If it is necessary to change a block library, either to add more blocks or to edit a block, the block library must be unlocked. To unlock a block library, choose **Edit:Unlock Library** from the block library window menu bar. The library will be unlocked as long as it remains open. To relock a block library, close the library, then reopen it.

When a block in a library is changed, the change will be applied to each corresponding reference block when the model containing the reference block is opened or run, or when **File:Update Diagram** is selected from the model window menu bar.

To quickly find the library block corresponding to a reference block, select the reference block, and then choose **Edit:Link Options:Go to Library Block** from the model window menu bar.

Unlinking a Block.   A reference block can be converted into a normal block by breaking the link to the library block. To remove the link, select the block, and then choose **Edit:Link Options:Disable Link** from the model window menu bar. Breaking a link affects only the instance of the reference block that was selected. The library block and all other corresponding reference blocks are unaffected.

Troubleshooting Links.   If Simulink is not able to find a library block corresponding to a reference block, the reference block is displayed with a red dashed border, and an error message is displayed. To correct the problem, delete the reference block from the model and then reinstall the reference block. An alternative is to double-click the reference block. A dialog box will appear, prompting you for the path to the library block.

■
EXAMPLE 7.11   Creating a custom block library

In this example, we will create a custom block library. The library will contain a Gain block and a block library containing the Spring-mass block.

Open a new library window by choosing **File:New:Library**.

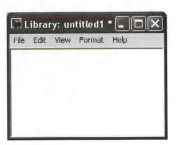

Copy a Gain block from the Math Operations block library and a Subsystem block from the Ports & Subsystems block library.

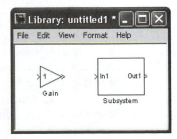

Label the subsystem My Library.

Double-click subsystem My Library, opening the subsystem window. Delete the Inport and Outport blocks and signal line, then drag a copy of the Spring-mass block to the subsystem window, and rename the block Spring-mass. Then, close the subsystem window. Select the subsystem and choose **Edit:Mask Subsystem** from the model window menu bar.

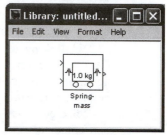

In the **Drawing commands** field on the **Icon** page, enter disp('Personal \nMasked\nBlocks') ;. Leave all other fields in the Mask Editor blank.

Choose **OK**. Resize My Library to fit the block icon. Save the model using the name PersonalLibrary. Then, close the subsystem window. Select the subsystem and choose **Edit:Mask Subsystem** from the model window menu bar.

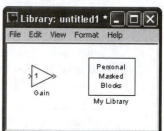

7.3.9 Configurable Subsystems

Configurable subsystems provide a convenient means of adding optional functionality to a model. For example, suppose we wish to implement the automobile model of Figure 7.1 such that the user can choose proportional control or proportional-derivative (PD) control. We could place both controllers in a Configurable Subsystem and then choose between the two controllers before running the simulation. To illustrate the procedure, we will build a configurable subsystem that has two inputs, (a and b) and a single output c. The subsystem is to be configurable to produce either $c = a + b$ or $c = k * a$, where k is a parameter.

The first step in building a configurable subsystem is to build a block library that contains blocks that provide the desired functions. In this case, the library contains two subsystems. The library and the subsystem blocks it contains are shown in Figure 7.28. Notice that subsystem a + b has two inputs, and subsystem k*a has only one input. k*a is masked and has a single prompt **Gain:** on the **Parameters** page and corresponding variable k. Save the block library after adding all subsystems.

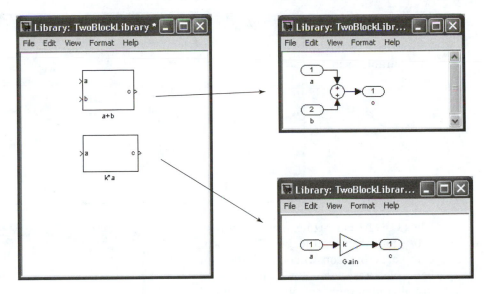

FIGURE 7.28: Block library for configurable subsystem

Drag a Configurable Subsystem block from the Signals & Systems block library to the new block library window.

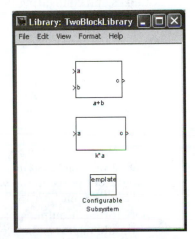

Double-click the Configurable Subsystem block. The Configuration dialog box will appear as shown. Notice that there is a check box for each subsystem in the block library.

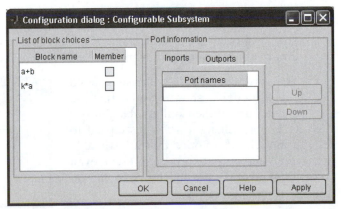

Click both check boxes to add the subsystems to the configurable subsystem.

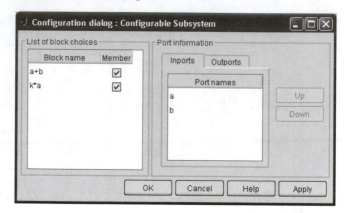

Click **OK**. After resizing, the Configurable subsystem displays the port labels. This completes the construction of the configurable subsystem. You can build additional configurable subsystems in a block library by adding Configurable Subsystem blocks and configuring them as needed.

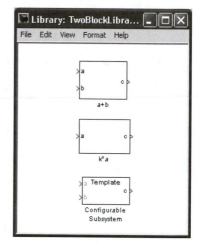

When you create a set of subsystems for a block library that will be used with configurable subsystems, you must pay special attention to the naming of the input and output ports on the various subsystems. The configurable subsystem block will have one input port for each input port name in the block library and one output port for each output port name. In this example, the same names were used for both subsystems in the block library. But if the output port on the second subsystem (k*a) were changed to d, the configurable subsystem block would have two output ports. Because a block cannot have input and output ports with the same name, an error will occur if the same name is used for an input in one subsystem in a block library and for an output in another subsystem in the same block library.

To use the configurable subsystem, build a new model as shown. Notice that the appearance of the Configurable Subsystem block changes when it's copied to a model. Also notice that all available input and output ports are present, even if not all of the possible block configurations use all of the ports.

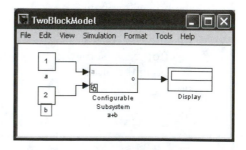

To choose a subsystem, select the Configurable Subsystem block, and choose **Edit:Block** choice. All available subsystems are listed in the drop-down list. The **Edit** menu also provides options for setting subsystem parameters and properties. If you double-click the Configurable Subsystem block, it will open the currently selected subsystem or, if masked, the masked block dialog box.

7.4 CONDITIONALLY EXECUTED SUBSYSTEMS

The Ports & Subsystems block library provides two blocks that cause subsystems to execute conditionally. The Enabled subsystem block causes a subsystem to execute only if a control input is positive. The Triggered subsystem block causes a subsystem to execute once when a trigger signal is received. Placing both the Enabled subsystem and Triggered subsystem blocks in a subsystem causes the subsystem to execute once when a trigger signal is received, only if an enable input is positive.

7.4.1 Enabled Subsystems

Enabled subsystems provide a means for modeling systems that have multiple operating modes or phases. For example, the aerodynamics of a jet fighter in the landing configuration are quite different from the aerodynamics of the same airplane in supersonic flight. The digital flight control system for such an airplane likely uses different control algorithms in the different flight regimes. A Simulink model of the airplane and its control system might need to include both flight regimes. It is possible to model such a system using logic blocks or switch blocks. However, that would require every block in the model to be evaluated each simulation time step, including blocks not currently contributing to the system's behavior. If we convert the various flight dynamics and control algorithm subsystems into enabled subsystems, only the subsystems that are active during a particular simulation step will be evaluated during that step. This can provide a significant computational savings.

A subsystem is converted into an enabled subsystem by adding an Enable block from the Ports & Subsystems block library to the subsystem. The Enabled Subsystem block provides a subsystem preconfigured with an Enable block. In Figure 7.29, a simple proportional controller converted into an enabled subsystem is illustrated.

The Enable block's dialog box is shown in Figure 7.30. The dialog box has three fields. The first field, **States when enabling**, is a drop-down menu with two options: **reset** and **held**. Choose **reset** to cause any internal states in the subsystem

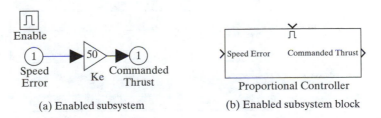

(a) Enabled subsystem   (b) Enabled subsystem block

FIGURE 7.29: Creating an enabled subsystem

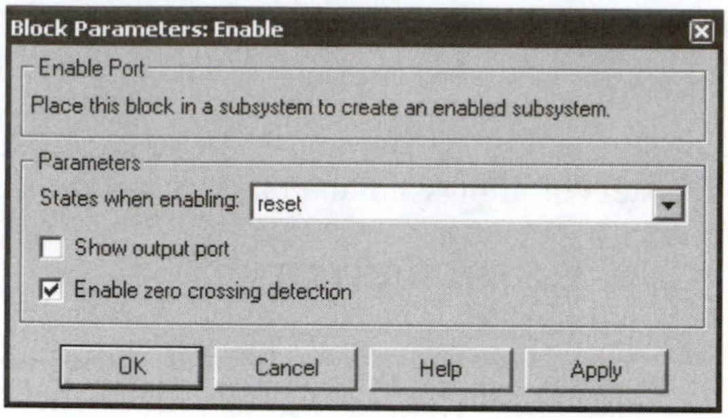

FIGURE 7.30: Enable block dialog box

to be reset to the specified initial conditions each time the block is enabled. If you choose **held**, when the block is reenabled, it will resume with all internal states at the values they held when the block was last executed. The second field, **Show output port**, is a check box. When selected, the Enable block will have an output port. This output port passes through the signal received at the Enable input port when the block is enabled. The third field, **Enable zero crossing detection** is a check box used to enable or disable zero crossing detection for the enable signal.

It is also important to configure the Outport blocks of an enabled subsystem. The dialog box for the Outport block (Figure 7.31) has three fields. The first field, **Port number**, determines the order in which the ports are displayed on the subsystem block icon. The second field, **Output when disabled**, is a drop-down menu with two options: **reset** and **held**. Choose **reset** to cause the output to reset to the value in the third field, **Initial output**. Choose **held** to cause the output to remain at the last value output before the subsystem was disabled.

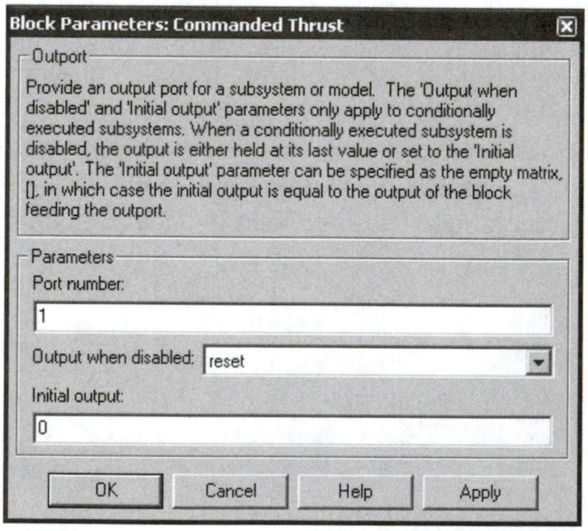

FIGURE 7.31: Enabled subsystem Outport block dialog box

An enabled subsystem is enabled when the signal at the Enable input port is positive. The input signal may be a scalar or a vector. If the signal is a vector, the subsystem is enabled if any element of the vector is positive.

EXAMPLE 7.12   Using an enabled subsystem

To illustrate the use of enabled subsystems, suppose we wish to modify the automobile speed control of Example 6.6 such that it has two modes of operation depending on the speed error. If the absolute value of speed error

$$v_{err} = |\dot{x}_{desired} - \dot{x}|$$

is less than a threshold value, say 2 ft/sec, and the absolute value of the rate of change speed error, $\dot{v}_{err}$, is also less than a threshold value, say 1 ft/sec^2, we wish to switch to proportional-integral (PI) control. Once PI control is enabled, it is to remain enabled as long as v_{err} is less than a larger threshold value, 5 ft/sec. Otherwise, proportional control is to be enabled.

The Simulink model is shown in Figure 7.32. The mode selector subsystem, shown in Figure 7.33, produces two outputs. Choose PI is set to 1.0 if the conditions for PI control are satisfied, and 0.0 otherwise. Choose P is always the logical inverse of Choose PI.

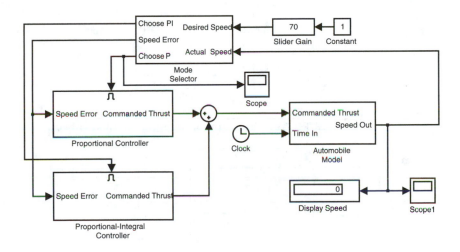

FIGURE 7.32: Simulink model of automobile with a dual-mode speed control

The proportional controller subsystem is illustrated in Figure 7.29. The PI subsystem block is shown in Figure 7.34. The Enable block and Outport block for both controller subsystems are configured to reset.

Executing the simulation results in the speed trajectory shown in Figure 7.35(a). The Choose P signal for this simulation is plotted in Figure 7.35(b). Initially, proportional control is enabled. Because the rate of change of speed error is less than the threshold value, as soon as the speed error decreases below 2 ft/sec, PI control is enabled.

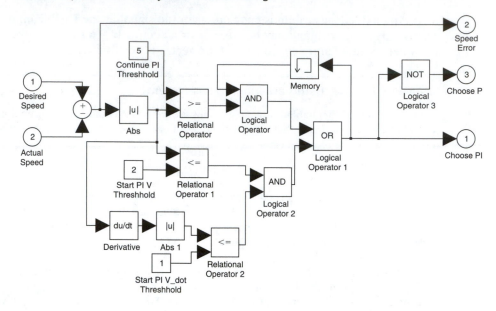

FIGURE 7.33: Mode selector subsystem

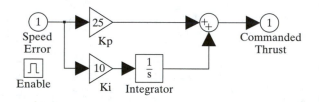

FIGURE 7.34: Proportional-integral enabled subsystem

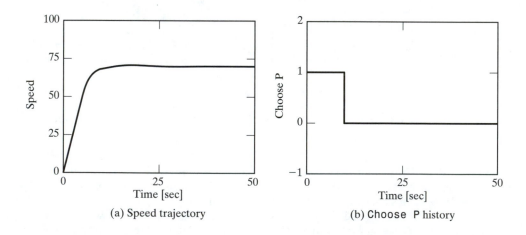

(a) Speed trajectory

(b) Choose P history

FIGURE 7.35: Speed trajectory and Choose P signal history for dual-mode controller model

7.4.2 Triggered Subsystems

A triggered subsystem is executed once each time a trigger signal is received. A triggered subsystem and the trigger dialog box are shown in Figure 7.36. The first field, **Trigger type**, is a drop-down menu with three choices: rising, falling, and either. If rising is selected, a trigger signal is defined as the trigger input crossing zero while increasing. If falling is selected, the trigger signal is defined as the trigger input crossing zero decreasing. If either is selected, the trigger signal is defined as the trigger input crossing zero increasing or decreasing. The second field in the Trigger block dialog box, **Show output port**, is a check box. If the box is checked, the Trigger block will have an output port that passes through the trigger signal. Field **Output data type** is a drop-down menu that is available only if an output port is enabled. The menu can be set to Auto, double, or int8. If Auto is selected (the default choice), the data type will be determined by the block to which the output port is connected. The final field is check box **Enable zero crossing detection**, which enables zero crossing detection for the selected block only.

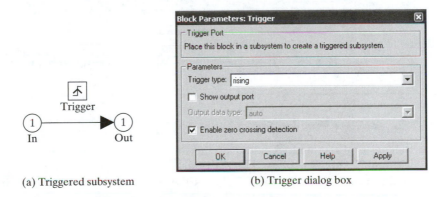

(a) Triggered subsystem   (b) Trigger dialog box

FIGURE 7.36: From Workspace block and dialog box

A triggered subsystem holds its output value after the trigger signal is received. The initial output value of a triggered subsystem is set using the subsystem's Outport blocks.

The trigger signal may be a scalar or a vector. If the signal is a vector, the subsystem is triggered when any element of the vector satisfies the **Trigger type selection**.

EXAMPLE 7.13   Using a triggered subsystem

The triggered subsystem illustrated in Figure 7.36 passes its input to its output when a trigger signal is received and holds its output at that value until another trigger signal is received. The Outport block is initialized to 0. In Figure 7.37, we added this subsystem to the automobile speed control model and routed the subsystem output to a Display block from the Sinks block library. The trigger input is connected to the enable signal for the PI controller. The Display block displays 0 until the PI controller is activated, and afterwards, it displays the most recent activation time.

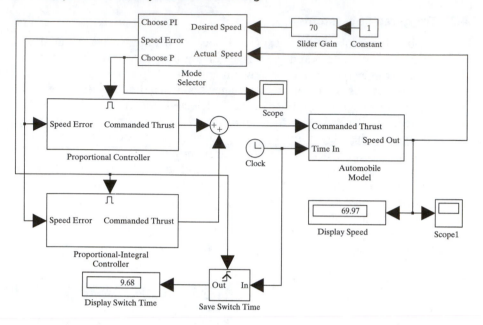

FIGURE 7.37: Automobile model with triggered subsystem

7.4.3 Trigger-When-Enabled Subsystems

Placing an Enable block and a Trigger block in a subsystem produces a triggered-when-enabled subsystem. This subsystem will have both a trigger input and an enable input. The subsystem behaves the same as a triggered subsystem, but the trigger signal is ignored unless the enable signal is positive.

7.4.4 Discrete Conditionally Executed Subsystems

All three types of conditionally executed subsystems may contain continuous blocks, discrete blocks, or continuous and discrete blocks. Discrete blocks in an enabled subsystem execute based on their sample time. They use the same time reference as the rest of the Simulink model; time in the subsystem is referenced to the start of the simulation and not the activation of the subsystem. Consequently, an output dependent upon a discrete block won't necessarily change at the instant the subsystem is enabled.

Discrete blocks in triggered subsystems must have their sample times set to −1, indicating that they inherit their sample time from the driving signal. Note that discrete blocks that include time delays (z^{-1}) change state once each time the subsystem is triggered.

7.5 LOGICAL SUBSYSTEMS

Simulink provides two types of logical conditional subsystems: if subsystems and switch-case subsystems. Both types of logical subsystems employ two blocks: test blocks and action subsystem blocks. The test blocks perform specified logical tests on one or more input signals and activate one action subsystem block depending on the test result. Thus, at any instant, at most one of the action subsystems associated with a test block is active, and the rest are dormant.

The action subsystems are standard subsystems with an additional action port and corresponding action port block. An action subsystem is built in a manner identical to an enabled subsystem.

EXAMPLE 7.14   Using an if subsystem

In this example, we will modify the dual-mode automobile speed control to use an if subsystem. Proportional control will be enabled if the speed error is more than five miles per hour, and PI control will be enabled otherwise. The top-level Simulink model is shown in Figure 7.38. Notice that there is one if block and two if action subsystem blocks. The block dialog box for the if block is shown in Figure 7.39.

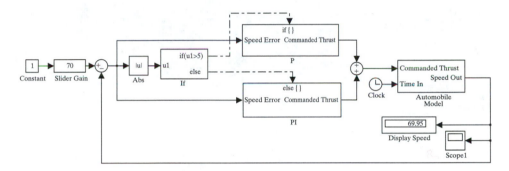

FIGURE 7.38: Dual-mode speed control with if subsystem

The proportional control if action subsystem is shown in Figure 7.40. Notice that the subsystem has an additional action port block.

7.6   ITERATIVE SUBSYSTEMS

Iterative subsystems provide a means with which to add simple control flow blocks to Simulink models. For complex control flow problems, Stateflow is more appropriate. There are two types of iterative subsystems: For Iterator subsystems and While Iterator subsystems. A For Iterator subsystem executes a specified number of iterations each time the subsystem is activated. A While Iterator subsystem executes repeatedly until the condition input is false. For Iterator subsystems and While Iterator subsystems are found in the Ports & Subsystems block library.

7.6.1   For Iterator Subsystem

Figure 7.41(a) shows the For Iterator block contained in the For Iterator Subsystem from the Ports & Subsystems block library. The For Iterator block controls the number of iterations each time the subsystem is invoked. The For Iterator block dialog box is shown in Figure 7.41(b). Set field **Source of number of iterations** to

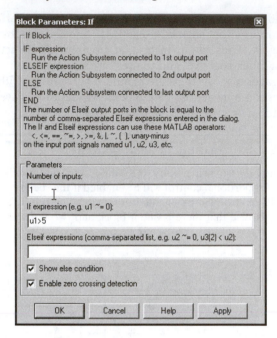

FIGURE 7.39: If block dialog box

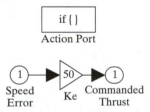

FIGURE 7.40: Proportional control if action subsystem

`internal` to specify the number of iterations in field **Number of iterations** or to `external` to add a number of iterations input port to the subsystem and For Iterator block.

7.6.2   While Iterator Subsystem

The While Iterator subsystem is similar to the For Iterator subsystem. The While Iterator block [Figure 7.42(a)] can be configured via the block dialog box [Figure 7.42(b)] to implement while behavior or do-while behavior. If while behavior is implemented, and the value at the initial condition port is zero, no iterations will be performed, and the block output will be as determined by the subsystem output port configuration. If the first iteration is executed, it will continue until the signal entering the cond input port is zero or the specified maximum number of iterations are complete. Selecting the do-while configuration forces the first iteration to occur.

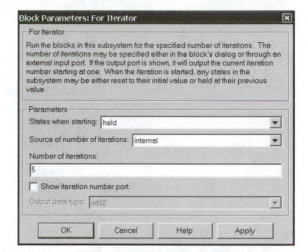

For
Iterator

For Iterator

(a) For Iterator Subsystem   (b) For Iterator block dialog box

FIGURE 7.41: For Iterator subsystem

cond while {
     ...
IC   }

While Iterator

(a) While Iterator Subsystem   (b) While Iterator block dialog box

FIGURE 7.42: While Iterator subsystem

EXAMPLE 7.15   Using the While Iterator Subsystem

To illustrate the use of iterative subsystems, suppose a block output is an implicit function of the block input.

$$yu + \exp y - 3 = 0$$

where u is the input signal, and y is the output signal. We can compute the output by solving the implicit function iteratively using a Newton–Raphson technique. (For details, see Section 13.3.1.) For each iteration, we compute the new estimate of the output y^+ as a function of the previous estimate y^- and the input:

$$y^+ = y^- - \left(y^- u + \exp y^- - 3\right) / \left(u + \exp y^-\right)$$

The Simulink model shown in Figure 7.43 implements this algorithm using a While Iterator subsystem.

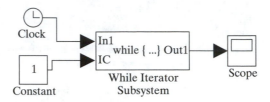

FIGURE 7.43: Simulink model with While Iterator Subsystem

The While Iterator subsystem (Figure 7.44) contains two Function blocks. Function block Update computes the next value of y using the expression

```
u(2) - (u(1)*u(2) + exp(u(2)) - 3)/(u(1) + exp(u(2)))
```

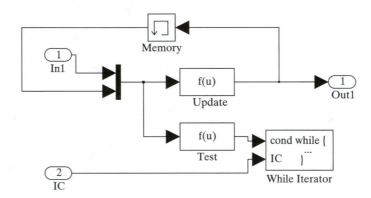

FIGURE 7.44: While Iterator subsystem

and Function block Test outputs a value of 1.0 until the error squared falls below a preset threshold (here 10^{-8}) using the expression

```
((u(1)* u(2) + exp(u(2)) - 3)^2) > 1E - 8
```

The purpose of the memory block in the iterative subsystem is to supply the result of the previous iteration (here y^-) to the current iteration. The memory block is initialized to a guess for the output for the initial invocation. For subsequent invocations, the initial value of Memory block output is the final value from the previous invocation.

The While Iterator block dialog box is shown in Figure 7.45. We set field **States when starting** to Held so that each time the Iterator subsystem is invoked, the starting guess for the block output is the final value from the previous invocation.

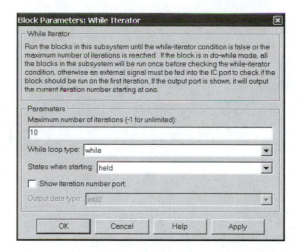

FIGURE 7.45: While Iterator block dialog box

7.7   SUMMARY

In this chapter, we described several Simulink features that make it practical to model complex systems. We discussed Simulink subsystems, which provide a facility similar to subprograms in traditional programming languages. We then discussed masking, which allows us to create subsystems that hide their functionality. Last, we discussed conditionally executed subsystems.

REFERENCE

[1] Hanselman, Duane C., and Littlefield, Bruce R., *Mastering MATLAB 6: A Comprehensive Tutorial* (Upper Saddle River, N.J.; Prentice Hall, 2001). This text provides a comprehensive tutorial on MATLAB programming.

8

Simulink Analysis Tools

In this chapter, you'll learn to use Simulink analysis tools to gain understanding of Simulink models and to aid in the design process. Using the linearization tools, we'll extract linear state-space models from block diagrams. We'll then use the trim tools to find equilibrium points. Finally, we'll use the Optimization Toolbox to optimize model parameters.

8.1 INTRODUCTION

In the previous chapters, we described the process of modeling dynamical systems using Simulink. In this chapter, we will show how you can use Simulink to gain insight into the behaviors of those systems.

All of the analysis and design capabilities we will discuss in this chapter are used from within the MATLAB workspace. From the MATLAB prompt, you can determine the structure of a model's state vector, the number of inputs and outputs, and other important parameters. You can run the simulation from within the MATLAB workspace using the MATLAB function `sim` and can change certain model parameters and inputs. The linearization commands (`linmod`, `linmod2`, and `dlinmod`) allow you to linearize a Simulink model about any point in its state-space. The trim command (`trim`) locates equilibrium points. These tools can be used to analyze the behavior of a system and to facilitate the design of certain system parameters.

8.2 DETERMINING MODEL CHARACTERISTICS

In order to use many of the capabilities of the analysis tools to be discussed in this chapter, you must know the structure of the Simulink state vector for the model. In this section, we will start with a brief discussion of Simulink state vectors. Then, we will show how to determine the structure of a model's state vector from the MATLAB command line.

8.2.1 Simulink State Vector Definition

A Simulink model is a graphical description of a set of differential, difference, and algebraic equations. Simulink converts this graphical representation into a state-space representation consisting of a set of simultaneous first-order differential and difference equations, and a set of algebraic equations. For example, the second-order continuous system shown in Figure 8.1 is represented by Simulink as two first-order

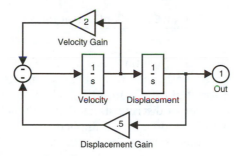

FIGURE 8.1: Simulink model of second-order continuous system

differential equations as follows:

$$\dot{x}_1 = x_2$$

$$\dot{x}_2 = -0.5x_1 - 2x_2 \tag{8.1}$$

Here, x_1 is a state variable corresponding to the output of the Integrator block labeled `Displacement`, and x_2 is a state variable corresponding to the output of the Integrator block labeled `Velocity`. Recall that the state variable representation of a system is not unique. An equally valid choice of state variables in this example would be to associate x_1 with `Velocity` and x_2 with `Displacement`. (Of course, if we change the state variable definition, we also must write the differential equations in terms of the new state variables.)

Because the analysis tools we will discuss in this chapter work in terms of the Simulink state vector, we must know how Simulink structures the state vector for a particular model. For example, the `sim` command includes the option to set the initial conditions for the model's state vector. If you wish to set the initial value of velocity to 1.0 and the initial value of displacement to 0.0, you must know which state variable corresponds to each quantity.

Simulink divides a model's state vector into continuous and discrete components. There is one continuous state vector component for each continuous integrator, including integrators present implicitly, due, for example, to Transfer Fcn blocks and State-Space blocks. Similarly, there is a discrete state vector component for each time delay, including delays present implicitly.

8.2.2   Using the model Command

The *model* command permits us to determine the structure of a Simulink model's state vector. There are three versions of the *model* command, as follows:

```
sizes = model([],[],[],0)

[sizes, x0] = model([],[],[],0)

[sizes, x0, states] = model([],[],[],0)
```

where *model* is the name of the Simulink model. Note that the input arguments are the same for each version. The first three arguments are empty matrices, and the fourth argument is 0.

TABLE 8.1: Contents of the `sizes` Vector

Component	Meaning
`sizes(1)`	Number of continuous states. This will include explicit states associated with Integrator blocks, and implicit states, such as those associated with Transfer Fcn blocks and State-Space blocks.
`sizes(2)`	Number of discrete states. This will include explicit states associated with Unit delays and implicit states, such as those associated with Discrete Transfer Fcn blocks.
`sizes(3)`	Number of outputs. An output is counted for each component of each Outport block. Thus, if the signal entering an Outport block is a scalar, there will be one output associated with that block. If the signal entering an Outport block is a three-component vector, there will be three outputs associated with that block. Note that To Workspace blocks To File blocks do not count as outputs.
`sizes(4)`	Number of inputs. An input is counted for each component of each Inport block. Note that From Workspace blocks and From File blocks do not count as inputs.
`sizes(5)`	Number of discontinuous roots in the system. This information does not pertain to the analysis tools discussed in this chapter.
`sizes(6)`	Flag that is set to 1 if a subsystem has direct feedthrough of an input. This information does not pertain to the analysis tools discussed in this chapter.
`sizes(7)`	Number of sample times. Continuous systems have one sample time; discrete and hybrid systems can have multiple sample times.

The first output argument, `sizes`, is a six-element vector defined in Table 8.1. The first two elements of `sizes` contain the number of continuous and discrete states. So, for example, if `sizes(1)` is 2 and `sizes(2)` is 3, the state vector will have a total of five elements. The first two elements will be the continuous states, and the last three elements will be the discrete states.

The second output argument (`x0`) is optional. If present, it is the initial value of the Simulink model's state vector. Recall that the integrator states may be initialized using the dialog box for each integrator, and that this initialization may be overridden using the **Workspace I/O** page of the **Simulation:Parameters** dialog box.

The third output argument (`states`) is a cell array that identifies the block associated with each component of the Simulink state vector using the convention

```
model file name/top level subsystem/2nd level
subsystem/.../block
```

So, for example, the state associated with an integrator block labeled `Velocity` at the top level of a model named `sysmdl_a` would be `sysmdl_a/Velocity`. If this same block were located in a subsystem named `subsys_1` in model `sysmdl_a`, the state would be named `sysmdl_a/subsys_1/Velocity`.

EXAMPLE 8.1   Determining model characteristics

Consider the Simulink model shown in Figure 8.2. Suppose we saved this model in a file named `SizesXmpA.mdl`.

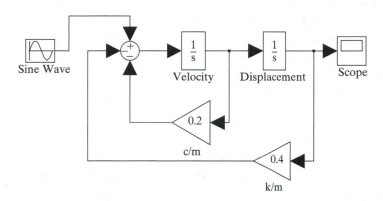

FIGURE 8.2: Second-order system with no inputs or outputs, stored in `SizesXmpA.mdl`

From the MATLAB prompt, we enter the following command:

```
[sizes, x0, states]=SizesXmpA([],[],[],0)
```

resulting in the output shown in Listing 8.1:

Listing 8.1: Model characteristics of `SizesXmpA.mdl`

```
>> [sizes, x0, states] = SizesXmpA([], [], [], 0)
sizes =
     2
     0
     0
     0
     0
     0
     1
x0 =
     0
     0
states =
    'SizesXmpA/Displacement'
    'SizesXmpA/Velocity'
```

From this output we can determine the structure of the model state vector. Component `sizes(1)` is 2, so there are two continuous states as we would expect, because the model has two Integrator blocks. Component `sizes(2)` is 0, so there are no discrete states. Using `sizes` and `states`, the state vector can be described as follows:

Component	Type	Associated with Block
sizes(1)	Continuous	Displacement
sizes(2)	Continuous	Velocity

EXAMPLE 8.2   Model characteristics with subsystems

We can also identify the model characteristics for a Simulink model containing subsystems. Suppose we saved the model shown in Figure 8.3 in file SizesXmpB.mdl. From the MATLAB prompt, we enter the command

```
[sizes, x0, states]=SizesXmpB([], [], [], 0)
```

resulting in the output shown in Listing 8.2.

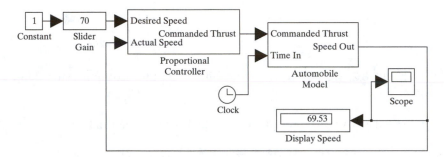

FIGURE 8.3: Automobile model using subsystems

Listing 8.2: Model characteristics of SizesXmpB

```
sizes =
     2
     0
     0
     0
     0
     0
     2
x0 =
     0
     0
states =
    'SizesXmpB/Automobile
Model/Speed'
    'SizesXmpB/Automobile
Model/Distance'
```

Note that the block names in states consist of the full path to the block, including the name of the subsystem.

8.3 EXECUTING MODELS FROM MATLAB

The MATLAB `sim` command permits us to run Simulink models from the MATLAB prompt or from within M-files. This capability makes it easy to access model states for analysis or to run a simulation repeatedly using different values for parameters, different inputs, or different initial conditions. Command `sim` may be used in conjunction with MATLAB command `simset`, which creates and edits a simulation options data structure, and `simget`, which obtains the simulation options structure from a model. After discussing these commands, we will show how they can be used to facilitate the analysis of model behavior.

8.3.1 Using `sim` to Run a Simulation

The syntax of the `sim` command is

```
[t,x,y] = sim(model,TimeSpan,Options,ut)
```

The return variable list is optional. It can be omitted, can contain only one return variable, two return variables, or all three return variables. The first returned variable (`t`) contains the value of simulation time at each output point. Recall that the default behavior is to produce an output point at the end of each integration step, but that the Output options section of the **Simulation:Parameters** dialog box allows you to override this default by adding output points or explicitly specifying the time of each output point. The second returned variable (`x`) is the state variable trajectory, with one column for each state variable, and one row corresponding to each time point in the first returned variable (`t`). The final returned variable (`y`) contains the output variable trajectory, one column for each output vector element, and one row corresponding to each time point. If there are no Outport blocks, but an output variable (`y`) is present, it will contain the empty vector [ ].

The first argument to the `sim` command (*model*) is a MATLAB string containing the name of the Simulink model, without the filename extension (`.mdl`). This argument is mandatory. The remaining arguments are optional and allow us to override various model configuration parameters.

`TimeSpan` specifies the output time points. If `TimeSpan` is specified, it overrides the output times specified in the **Simulation:Parameters** dialog box. `TimeSpan` may have four different forms, as listed in Table 8.2.

The MATLAB data structure `options` allows you to override many of the parameters in the **Simulation:Parameters** dialog box. The `options` structure is created, updated, and displayed using the MATLAB command `simset`, which we will discuss shortly.

`ut` overrides the Load from workspace section of the **Workspace I/O** page of the **Simulation:Parameters** dialog box. The `ut` may be either an input table or a string containing the name of a MATLAB function. If `ut` is an input table, it must be of the form [`t,u1,u2,...`], where `t` is a column vector of time points, and `u1`, `u2`, etc., are column vectors of the values of the inputs corresponding to the time points in `t`. The inputs are associated with Inport blocks, ordered according to the Inport block numbers. Simulink treats `ut` as a lookup table, linearly interpolating between time points. If `ut` is a string, it must name a function that returns an input vector, given a single argument of time.

TABLE 8.2: Specifying the Simulation Output Times in the `sim` Command

Value of `TimeSpan`	Output Time Point Values
`[ ]`	Default to values specified in the **Simulation:Parameters** dialog box.
`[T_Final]`	Time points default to values specified in the **Simulation:Parameters** dialog box. The simulation will stop when the simulation time reaches `T_Final`.
`[T_Start T_Final]`	The simulation will start at time `T_Start`, and stop at `T_Final`. The value of time at intermediate points will be as specified in the **Simulation:Parameters** dialog box. Note that this is a two-element vector.
`[OutputTimes]`	If `OutputTimes` is a vector of three or more components, there will be one output time point corresponding to each element of `OutputTimes`. The contents of the first return variable (`t`) will be identical to `OutputTimes`. Probably the most common form of `OutputTimes` is `[T_Start:TimeSpacing:T_final]`, producing a vector of equally spaced points.

When input arguments are supplied to the `sim` command to override simulation parameters, the Simulink model is not changed. The `sim` command arguments affect the Simulink model only during the execution of the simulation. When the simulation stops, the model parameters are the same as they were before the `sim` command was executed.

EXAMPLE 8.3   Creating an input table

Suppose we wish to set `ut` for a model with a single input according to the rule

$$u(t) = \frac{1}{2}t, \qquad\qquad t \le 1$$

$$u(t) = t - \frac{1}{2}, \qquad\qquad t > 1$$

The following MATLAB statement will create an input table compatible with the `sim` command. Simulink will interpolate in this table, producing values that correspond to the rule.

```
ut=[[0,1,100]',[0,1/2,99.5]']
```

EXAMPLE 8.4   Creating a vector input table

Suppose we wish to set `ut` for a model with two inputs to the vector function

$$\mathbf{u}(t) = [\sin t \ \cos t]$$

The function M-file shown in Listing 8.3 returns this vector:

Listing 8.3: M-file to return the vector [sin(t),cos(t)]

```
function u = ut_fun(t)
u = [sin(t),cos(t)] ;
```

We set ut using the MATLAB statement

```
ut='ut_fun'
```

EXAMPLE 8.5   Running a model using the sim command

In this example, we run the model illustrated in Figure 8.2 using the model's default configuration. This is equivalent to selecting **Simulation:Start** from the model's menu bar. At the MATLAB prompt, enter the command:

```
sim('SizesXmpA')
```

Simulink will run the model, and if the Scope is open, will display the plot of the output when the simulation is complete.

8.4   SETTING SIMULATION PARAMETERS WITH SIMSET

The simset command is used to create and edit the options structure. There are three forms of the simset command. The first form is

```
options = simset('RelTol',1.0E-4,'Solver','ode4')
```

The second form of simset is

```
options = simset(oldopts,'name_1',value_1,...)
```

which allows you to change an existing options structure, either by changing previously set options or by adding more options. The third form of simset is

```
options = simset(oldopts,newopts)
```

where both oldopts and newopts are options structures previously created using simset. This form of simset merges the contents of oldopts and newopts, with the contents of newopts having priority.

Table 8.3 describes the properties that may be set using simset.

It is particularly important to ensure that the properties SrcWorkspace and DstWorkspace are set properly. There are three options for each: 'base', 'current', and 'parent'. The base workspace is the MATLAB prompt. Thus, variables defined in the base workspace are those variables listed when the command who is entered at the MATLAB prompt. Each function M-file has a private

TABLE 8.3: Simulation Property Names and Values

Property Name	Property Value and Associated **Simulation:Parameters** Fields
Solver	Selects the Simulink differential equation solver. Permissible values are: `'ode45'`, `'ode23'`, `'ode113'`, `'ode15s'`, `'ode23s'`, `'ode5'`, `'ode4'`, `'ode3'`, `'ode2'`, `'ode1'`, `'FixedStepDiscrete'`, `'VariableStepDiscrete'`.
RelTol	Overrides the contents of the **Relative tolerance** field.
AbsTol	Overrides the contents of the **Absolute tolerance** field.
Refine	Equivalent to setting **Output options** to `Refine` output. The property value must be set to a positive integer. If the output time points are specified explicitly, this property is ignored.
MaxStep	Upper bound on integration step size. Overrides **Max step size**.
InitialStep	Initial integration step size. Overrides the contents of **Initial step size**.
MaxOrder	Maximum order if solver ODE15S is used. Otherwise ignored.
FixedStep	Integration step size if a fixed step solver is used. Overrides **Fixed step size**.
OutputPoints	Set to either the string `'specified'` or the string `'all'`. Select `'specified'` (the default) to produce output points only at the time points specified in `TimeSpan`. Select `'all'` to produce output points at the time points specified in `TimeSpan` and at each integration step. This property is equivalent to selecting `Produce specified output only` or `Produce additional output` in **Output options**.
OutputVariables	Overrides the three check boxes in the Save to workspace section of the **Workspace I/O** page. The property value is a string of up to three characters. If the string contains `'t'`, simulation time is output to the MATLAB workspace, equivalent to selecting the **Time** check box. Including `'x'` is equivalent to selecting the **States** check box, and including `'y'` is equivalent to selecting the **Output** check box. Thus, this property value could be set to `'txy'`, `'ty'`, `'yx'`, etc.
MaxRows	Places an upper bound on the number of rows in the output matrices. Equivalent to selecting the **Limit to last** check box and entering a value in the **Limit to last** input field.
Decimation	Overrides the **Decimation** field. Setting this property value to 1 causes every point to be output, setting it to 2 causes every other point to be output, and so on. Must be a positive integer.

TABLE 8.3: Simulation Property Names and Values (Cont)

Property Name	Property Value and Associated **Simulation:Parameters** Fields
`InitialState`	Set to a vector containing the initial values of all of the model's state variables. Overrides the **Load initial** check box and field in the States section of the **Workspace I/O** page.
`FinalStateName`	Character string containing the name of the MATLAB workspace variable in which to save the final value of the state vector. Overrides the **Save final** check box and field in the States section of the **Workspace I/O** page.
`Trace`	Set to contain a comma-separated list of strings that may include `'minstep'`, `'siminfo'`, `'compile'`, or `''`. For example, to include all three options, the string could be `'siminfo,compile,minstep'`. `'minstep'` is equivalent to setting the **Minimum step size** violation event to `warning`. `'siminfo'` causes Simulink to produce a brief report listing key parameters in effect at the start of a simulation. `'compile'` causes Simulink to produce a model compilation listing in the MATLAB workspace as the model is compiled, before it runs. The compilation listing is intended primarily for use by The MathWorks in troubleshooting Simulink problems.
`SrcWorkspace`	Select the MATLAB workspace in which to evaluate MATLAB expressions defined in the model. May be set to `'base'` (the default), `'current'`, or `'parent'`. This option is extremely important if you are running a Simulink model from within a function M-file. `'base'` indicates that all variables used to initialize block parameters are to be taken from the base MATLAB workspace. `'current'` indicates that the variables are to be taken from the private workspace from which `sim` is called. `'parent'` indicates that the variables are to be taken from the workspace from which the current function (the one that called `sim`) was called (the next higher function). There is no corresponding field in **Simulation:Parameters**.
`DstWorkspace`	Select the MATLAB workspace in which to assign MATLAB variables defined in the model. May be set to `'base'`, `'current'`, or `'parent'`. `DstWorkspace` determines where To Workspace blocks send their results. There is no corresponding field in **Simulation:Parameters**.
`ZeroCross`	Overrides the **Disable zero crossing detection** check box. May be either `'on'` or `'off'`.

workspace; variables defined in a function M-file are not visible outside the M-file. 'current' indicates that variables are to be taken from or returned to the workspace from which the sim command is run. 'parent' indicates that variables are to be taken from or returned to the workspace from which the currently executing function was called. It is permissible for SrcWorkspace and DstWorkspace to be set to different values.

8.4.1 Obtaining Simulation Parameters with simget

The command simget gets either the options structure or the value of a single property. To get the options structure from a model, the syntax is

    opts = simget(model)

To get the value of a single property, the syntax is

    value = simget(model,property_name)

where model is a MATLAB string containing the name of the Simulink model without the filename extension (.mdl), and property_name is a string containing one of the names listed in Table 8.3. property_name can also be a cell array in which each cell is a string containing one of the names listed in Table 8.3.

EXAMPLE 8.6   Drawing a simple state portrait

A useful tool in the analysis of nonlinear second-order systems is the state portrait. Consider the following system, discussed in detail by Scheinerman [6]:

$$\dot{x}_1 = -x_2$$
$$\dot{x}_2 = x_1 + x_2^3 - 3x_2$$

A state portrait graphically depicts the behavior of the system in the state plane [the (x_1, x_2) plane] in the vicinity of an *equilibrium*. An equilibrium is a point in the state plane at which both state derivatives are zero. If, in some sufficiently small region in the vicinity of an equilibrium, every trajectory converges to the equilibrium, the equilibrium is said to be stable. This system has one equilibrium, located at the origin. A Simulink model of this system is shown in Figure 8.4.

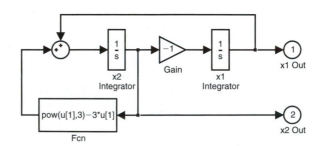

FIGURE 8.4: Simulink model of nonlinear second-order system

We start the analysis process by identifying the elements of the model's state vector. In Listing 8.4, we execute the model, stored in file `StatePrtXmp.mdl`, with the fourth argument (`flag`) set to 0. We note that there are two states, the first associated with x_1 and the second with x_2.

Listing 8.4: Identifying the states from the MATLAB prompt

```
>> [sizes, x0, states] = StatePrtXmp([], [], [], 0)
sizes =
     2
     0
     2
     0
     0
     0
     1
x0 =
     0
     0
states =
     'StatePrtXmp/x1
Integrator'
     'StatePrtXmp/x2
Integrator'
```

We can produce a simple state portrait by repeatedly running the model, starting each simulation at a different point in the state space. An M-file that produces a state portrait in this manner is shown in Listing 8.5. The state portrait is shown in Figure 8.5. Notice that each state trajectory goes to the origin and stops there. Thus, from the state portrait, it appears that the origin is a stable equilibrium.

Listing 8.5: M-file to produce simple state portrait

```
% Produce a state portrait for the SIMULINK model
% StatePrtXmp
%
for x1 = -1:0.5:1
   for x2 = -1:0.5:1
      opts = simset('InitialState',[x1,x2]) ;
      [t,x,y] = sim('StatePrtXmp',15,opts) ;
      hold on ;
      plot(x(:,1),x(:,2)) ;
   end
end set(gca,'FontSize',14) ;
axis([-1,1,-1,1]);
set(gca,'Ytick',[-1,-0.5,0,0.5,1]) ;
set(gca,'Xtick',[-1,-0.5,0,0.5,1]) ;
xlabel('x1') ;
   ylabel('x2') ;
```

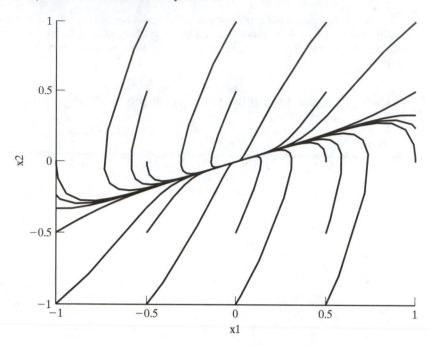

FIGURE 8.5: Simple state portrait

EXAMPLE 8.7 Quiver plot state portrait

The simple state portrait in the previous example shows the system behavior close to the origin and is easy to produce, but there are some problems with this approach. First, there is no easy method for determining the proper TimeSpan. Some of the trajectories approach the origin rapidly, others slowly. A more serious problem is that if the simulation is started at a point in the state space where the system is unstable, one or more of the state variables will rapidly diverge, causing floating-point overflow and failure of the Simulink simulation. This is not a weakness of Simulink; it is the natural consequence of modeling the behavior of an unstable system.

The M-file displayed in Listing 8.6 overcomes these problems. As before, we run the simulation repeatedly, starting each simulation at a different point in the state space. However, each simulation performs only one integration step. Using the final point of each state trajectory and the corresponding initial point, we form an approximation of the slope of the state portrait at each initial point. Using these slopes, we produce the quiver plot shown in Figure 8.6. In addition to being easier to read, this state portrait shows the behavior of the system in the unstable region of the state plane.

Listing 8.6: M-file to produce quiver plot state portrait

```
% Produce a state portrait for the SIMULINK model StatePrtXmp
% using a quiver plot
h = 0.01 ; % Fixed step size
opts = simset('Solver','ode5','FixedStep',h) ;
```

```
x1 = -2.5:0.25:2.5 ;    % Set up the x1 and x2 grid
x2 = -2.5:0.25:2.5 ;
[nr,nc] = size(x1) ;
x1m = zeros(nc,nc) ;  % Allocate space for the output
                      % vectors
x2m = x1m ;
for nx1 = 1:nc
  for nx2 = 1:nc
    opts = simset(opts,'InitialState',[x1(nx1),x2(nx2)]) ;
    [t,x,y] = sim('StatePrtXmp',h,opts) ;
    dx1 = x(2,1)-x1(nx1) ;
    dx2 = x(2,2)-x2(nx2) ;
    l = sqrt(dx1^2 + dx2^2)*7.5 ;  % Scale the arrows
    if l > 1.e-10
      x1m(nx2,nx1)=dx1/l ;            % Notice the reversed
                                     % indexes

      x2m(nx2,nx1)=dx2/l ;
    end
  end
end
quiver(x1,x2,x1m,x2m,0) ;
set(gca,'FontSize',14) ;
axis([-2.5,2.5,-2.5,2.5]) ;
set(gca,'Ytick',[-2, -1, 0, 1, 2]) ;
set(gca,'Xtick',[-2, -1, 0, 1, 2]) ;
xlabel('x1') ;
ylabel('x2') ;
```

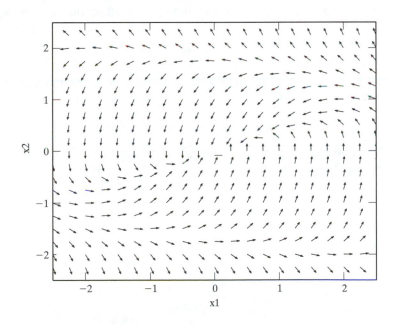

FIGURE 8.6: Quiver plot state portrait

8.5 LINEARIZATION TOOLS

While the dynamics of most physical systems are nonlinear, many useful techniques for analysis and control system design rely on linear models. For example, frequency domain analysis tools such as Bode plots and root locus plots are based on linear systems theory. Modern control techniques such as pole placement via state feedback and the linear quadratic regulator also rely on linear systems theory. Thus, it is frequently convenient to form linear approximations to nonlinear systems. We will begin this section with a short discussion of linearization. Then, we will show how the Simulink linearization tools may be used to facilitate analysis of nonlinear systems and control system design.

8.5.1 Linearization

A linear model of a nonlinear system is formed by computing matrices of the partial derivatives of the system state vector time rate of change and the output vector, with respect to the state vector and input vector. As discussed in Chapter 2, the general form of the state-space model of a dynamical system is

$$\dot{\mathbf{x}} = \mathbf{f}(\mathbf{x}, \mathbf{u}, t)$$

$$\mathbf{y} = \mathbf{g}(\mathbf{x}, \mathbf{u}, t)$$

where $\mathbf{x}$ is a state vector of order n, $\mathbf{u}$ is an input vector of order m, and $\mathbf{y}$ is an output vector of order p. Given a nominal point $\mathbf{x}$ and a nominal input $\mathbf{u}$, at some time t, we can approximate to first-order the change in the jth component of the state vector time rate of change ($\dot{x}_j$) due to a small variation in the ith component of the state vector (x_i) as

$$\delta \dot{x}_{ji} = \frac{\partial}{\partial x_i} f_j(\mathbf{x}, \mathbf{u}, t) \delta x_i$$

and the change due to a small variation in the kth component of the input vector as

$$\delta \dot{x}_{jk} = \frac{\partial}{\partial u_k} f_j(\mathbf{x}, \mathbf{u}, t) \delta u_k$$

The total change in $\dot{\mathbf{x}}$ due to a small variation in each element of the state vector and input vector can be approximated by the matrix equation

$$\delta \dot{\mathbf{x}} = \mathbf{A} \, \delta \mathbf{x} + \mathbf{B} \, \delta \mathbf{u}$$

where each element of $\mathbf{A}$ is

$$A_{ji} = \frac{\partial}{\partial x_i} f_j(\mathbf{x}, \mathbf{u}, t) \tag{8.2}$$

and each element of $\mathbf{B}$ is

$$B_{jk} = \frac{\partial}{\partial u_k} f_j(\mathbf{x}, \mathbf{u}, t)$$

Note that $\mathbf{A}$ is composed of n rows and n columns, and $\mathbf{B}$ is composed of n rows and m columns.

Using similar reasoning, we can approximate the change in the output vector due to small variations in the state and input as

$$\delta \mathbf{y} = \mathbf{C}\,\delta \mathbf{x} + \mathbf{D}\,\delta \mathbf{u}$$

where

$$C_{ji} = \frac{\partial}{\partial x_i} g_j(\mathbf{x}, \mathbf{u}, t)$$

and

$$D_{jk} = \frac{\partial}{\partial u_k} g_j(\mathbf{x}, \mathbf{u}, t)$$

Here, **C** is composed of p rows and n columns, and **D** is composed of p rows and m columns.

If the nominal point is the origin (perhaps after a change of variables), matrices **A**, **B**, **C**, and **D** may be used to form the linear state-space approximation:

$$\dot{\mathbf{x}} = \mathbf{A}\mathbf{x} + \mathbf{B}\mathbf{u}$$

$$\mathbf{y} = \mathbf{C}\mathbf{x} + \mathbf{D}\mathbf{u}$$

This approximation may be valid only in some small region in the vicinity of the origin. If the system is time varying, the approximation may be valid for a short period of time. Furthermore, some nonlinear systems cannot be effectively linearized, because important features of the dynamics are lost in the linearization. However, for a large number of physical systems, the linear approximation is useful, providing insight into the behavior of the system and providing access to a variety of control system design techniques.

8.5.2 Simulink Linearization Commands

Simulink provides three commands to extract linear state-space approximations from Simulink models. linmod forms state-space models of continuous systems. linmod2 is an alternative to linmod that uses different techniques to obtain the linearization, striving to reduce truncation error. dlinmod forms linear approximations of systems containing continuous and discrete components.

The syntax of the linearization commands is as follows:

```
[A,B,C,D]=linmod(model,X,U,para,xpert,upert)

[A,B,C,D]=linmod2(model,x,u,para,apert,bpert,cpert,dpert)

[A,B,C,D]=dlinmod(model,ts,x,u,para,xpert,upert)
```

All of the output arguments and all of the input arguments, except model, are optional. If one output argument or no output arguments are specified, a MATLAB structure containing the four output matrices, the state, output, and input names (in cell arrays), and the operating point is returned. The input arguments are defined in Table 8.4. The output arguments are the four state-space matrices, as discussed in Section 8.5.1.

TABLE 8.4: Linearization Command Input Arguments

Argument	Definition
model	MATLAB string containing the name of the Simulink model. For example, if the Simulink model is stored in file ex1.mdl, the first argument would be 'ex1'. This argument is required.
x	Value of the model nominal state vector. Simulink will linearize the model about any point in the model's state-space. Identify the components of the model's state vector using the procedure described in Section 8.2. x defaults to a zero vector of appropriate dimension if either the argument x is not present or if it is the empty matrix [ ].
u	Value of the nominal input vector. Simulink will linearize input matrix and direct transmittance matrix about the specified input vector and state vector. The components of u correspond to the numbering of Inport blocks.
para	Vector containing three elements. The first is a perturbation value to be used in computing the numerical partial derivatives, and it defaults to 10^{-5} for linmod and dlinmod. For linmod2, para(1) is the minimum perturbation value that defaults to 10^{-8}. The second element is the time at which to perform the linearization, and it defaults to 0 sec. Set the third element to 1 to eliminate states associated with blocks not in the path from input to output.
xpert	Vector of perturbation values for each component of the system state vector. This vector of perturbations overrides the default, computed according to the rule Xpert=para(1)+1e-3*para(1)*abs(x).
upert	Vector of perturbation values for each component of the input vector. This vector overrides the default, according to the rule upert=para(1)+1e-3*para(1)*abs(u).
apert	Matrix of maximum perturbation values for the system matrix. apert must be either empty ([ ]) or the same size as A. linmod2 uses a fairly complex algorithm to compute the optimal perturbation value for each element of each matrix. The perturbation value for each element of the system matrix (A(i,j)) will be bounded from above by apert(i,j) and from below by para(1).
bpert	Matrix of maximum perturbation values for the input matrix (B). The perturbation value for each element of B will be bounded from above by the corresponding element of bpert and from below by para(1).
cpert	Matrix of maximum perturbation values for the output matrix (C). The perturbation value for each element of C will be bounded from above by the corresponding element of cpert, and it will be bounded from below by para(1).
dpert	Matrix of maximum perturbation values for the direct transmittance matrix (D). The perturbation value for each element of D will be bounded from above by the corresponding element of dpert and from below by para(1).
ts	Sample time for dlinmod. This defaults to the maximum sample time for the model.

The steps for using the linearization tools are as follows:

1. Prepare a Simulink model configured with Inport blocks for the inputs and Outport blocks for the outputs. Note that blocks from the Sources block library do not count as inputs, and blocks from the Sinks block library do not count as outputs. It is not necessary to have an Inport block, but there should be at least one Outport block.

2. Use the procedures discussed in Section 8.2 to determine the number and ordering of the model's states.

3. Execute the appropriate linearization command from the MATLAB command line or from within an M-file. If you are interested only in the system matrix, use a single output argument, such as

```
A = linmod('sysmdl_a')
```

If you need the entire linear state-space model, use all of the output arguments.

The Derivative block and Transport Delay block should not be used in models to be linearized, as they can cause numerical trouble for the linearization functions. Replace Derivative blocks with Switched Derivative blocks. The Switched Derivative block is the Linearization block library, which is, in turn, found in the Simulink Extras library.

Once you obtain the linear model, you can use a variety of MATLAB commands to analyze the system dynamics or to design linear controllers.

■

EXAMPLE 8.8   Linearizing a model

Consider the nonlinear system discussed in Example 8.6:

$$\dot{x}_1 = -x_2$$
$$\dot{x}_2 = x_1 + x_2^3 - 3x_2$$

We wish to linearize this system about the origin. Applying Equation (8.2), we compute

$$\mathbf{A} = \begin{bmatrix} 0 & -1 \\ 1 & -3 \end{bmatrix}$$

A Simulink model of this system is shown in Figure 8.4. From the state portrait (Figure 8.6), it appears that this system exhibits stable behavior without oscillation. Thus, we would expect both eigenvalues to be real and negative. Listing 8.7 illustrates checking eigenvalues with MATLAB. First, the system matrix is determined using linmod, then eig is used to compute the eigenvalues of the system matrix, which are real and negative.

Listing 8.7: Using `linmod` to compute the system matrix of a nonlinear second-order system

```
>> [A,B,C,D] = linmod('StatePrtXmp');
>> disp(A)
        0    -1.0000
   1.0000    -3.0000
>> disp(eig(A))
  -0.3820
  -2.6180
```

EXAMPLE 8.9   Designing a linear controller

In this example, we will discuss the design of a linear controller for a nonlinear system. Consider the cart with an inverted pendulum illustrated in Figure 8.7. This system approximates the dynamics of a rocket immediately after liftoff. The objective of the rocket control problem is to maintain the rocket in a vertical attitude while it accelerates. The objective in the control of this model is to move the cart to a specified position (x) while maintaining the pendulum vertical. Ogata [4] presents a detailed explanation of the design of a linear controller for this system using the technique of pole placement and MATLAB.

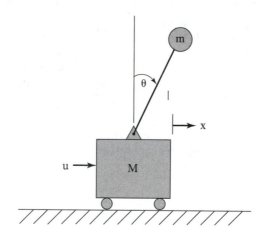

FIGURE 8.7: Cart with inverted pendulum

Referring to Figure 8.7, the input to the system is a horizontal force (u) applied to the cart of mass M. The pendulum is free to rotate without friction in the plane of the page. The pendulum is of length l, and the pendulum mass (m) is assumed to be concentrated at the end. The equations of motion of this system in terms of the cart displacement and pendulum angle can be written

$$(M + m)\ddot{x} - m l \dot{\theta}^2 + m l \ddot{\theta} \cos \theta = u$$

$$m \ddot{x} \cos \theta + m l \ddot{\theta} = m g \sin \theta$$

where g is the acceleration due to gravity. Examining these equations, we notice that $\ddot{x}$ and $\ddot{\theta}$ appear in both equations. If we build a Simulink model using these equations, the model will contain an algebraic loop. (See Chapter 13 for a detailed discussion of algebraic loops.) Employing a little algebra, the equations of motion can be rewritten to eliminate the algebraic loop:

$$\ddot{x} = \frac{m\,l\,\dot{\theta}^2 \sin\theta - m\,g\,\sin\theta\,\cos\theta + u}{M + m\sin^2\theta}$$

$$\ddot{\theta} = \frac{-(m\,l\,\dot{\theta}^2 \sin\theta\,\cos\theta - m\,g\,\sin\theta + u\,\cos\theta - M\,g\,\sin\theta)}{l(M + m\,\sin^2\theta)}$$

In order to follow the procedure of Ogata [4], our next task is to build a Simulink model of the system, with u as the input and x as the output. We choose as state variables $x, \dot{x}, \theta, \dot{\theta}$, and will need all four state variables in our controller. The model is enclosed in a subsystem with a scalar Inport block and a vector Outport block, as shown in Figure 8.8. Note that the equations of motion are computed using Function blocks, and the system parameters are variables defined in the MATLAB workspace. The subsystem is placed in a Simulink model that has a single input (u) and a single output (x), as shown in Figure 8.9. A Demultiplexer block decomposes the state vector into its components. x is routed to an Outport block, and the remaining components of the state vector are routed to Terminator blocks.

The system parameters are as follows:

$$M = 2\,\text{kg}$$

$$m = 0.5\,\text{kg}$$

$$l = 0.5\,\text{m}$$

$$g = 9.8\,\text{m/sec}^2$$

The M-file in Listing 8.8 determines the model characteristics, producing the output in Listing 8.9.

Listing 8.8: M-file to determine the characteristics of the inverted pendulum model

```
% Initialize the system parameters and obtain the
% model characteristics
%
% System parameters
l = 0.5 ;    % Pendulum length
g = 9.8 ;    % Gravity
m = 0.1 ;    % Pendulum bob mass
M = 2 ;      % Cart mass
% Get the model characteristics
[sizes,x0,states] = InvertPend([],[],[],0)
```

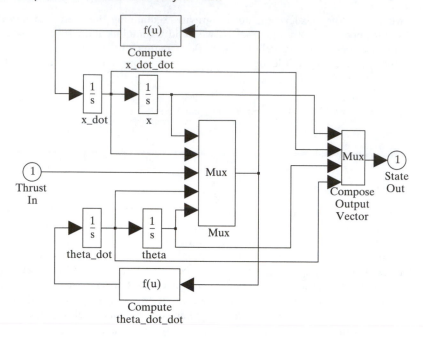

FIGURE 8.8: Inverted pendulum subsystem

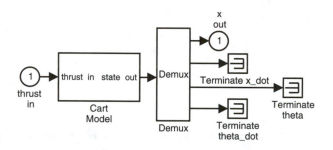

FIGURE 8.9: Simulink model to be linearized

Listing 8.9: Inverted pendulum model characteristics

```
sizes =
      4
      0
      1
      1
      0
      1
      1
x0 =
      0
      0
```

```
        0
        0
states =
      'InvertPend/Cart
Model/x'
      'InvertPend/Cart
Model/x_dot'
      'InvertPend/Cart
Model/theta_dot'
      'InvertPend/Cart
Model/theta'
```

Note that the Simulink state vector is

$$\begin{bmatrix} x \\ \dot{x} \\ \dot{\theta} \\ \theta \end{bmatrix}$$

which is ordered differently from the state vector at the outport block of the subsystem.

Next, we incorporate the subsystem in a Simulink model that includes a state feedback controller with an integrator in the forward path, shown in Figure 8.10. The input to the control system is the desired cart position. The Step function block is configured to produce a unit step at 1.0 sec. The output is the current cart position, displayed using a Scope block.

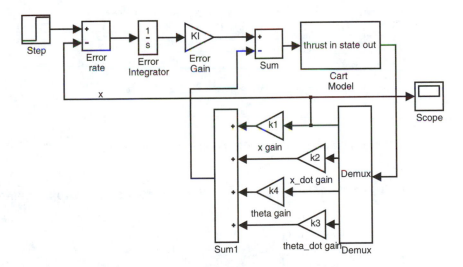

FIGURE 8.10: Inverted pendulum control Simulink model

Shown in Listing 8.10 is an M-file that computes the controller gains according to the procedure of Ogata [4]. The M-file uses linmod to obtain the state-space matrices. Then, Ackermann's method is used to compute the gains.

Listing 8.10: Controller design M-file

```
% Design the feedback gain matrix for the cart with inverted
% pendulum example. This example follows the procedure
% outlined in Example 2-3 of "Designing Linear Control
% Systems with MATLAB", by K. Ogata
%
% Note that for this model, the state vector is
% [x ; x_dot ; theta_dot ; theta]', so the arrangement
% of the system matrices will be different than in the
% Ogata example.
%
% System parameters
l = 0.5 ;    % Pendulum length
g = 9.8 ;    % Gravity
m = 0.1 ;    % Pendulum bob mass
M = 2 ;      % Cart mass
% Get the system matrices
[A,B,C,D] = linmod('InvertPend',[0,0,0,0],[0])
A1 = [A zeros(4,1) ; -C 0] ;
B1 = [B ; 0] ;
% Define the controllability matrix M
MM = [B1 A1*B1 A1^2*B1 A1^3*B1 A1^4*B1] ;
% Check the rank of MM, must be 5 if system in completely
% controllable
rank(MM)
% Obtain the characteristic polynomial
J = [-1+sqrt(3)*i,0,0,0,0;0,-1-sqrt(3)*i,0,0,0;...
    0,0,-5,0,0;0,0,0,-5,0;0,0,0,0,-5] ;
Phi = polyvalm(poly(J),A1) ;
% Get the feedback matrix using Ackermann's formula
KK = [0,0,0,0,1]*(inv(MM))*Phi;
k1 = KK(1)   ; % x gain
k2 = KK(2)   ; % x_dot gain
k3 = KK(3)   ; % theta_dot gain
k4 = KK(4)   ; % theta gain
KI = -KK(5)  ; % error integrator gain
```

The results of executing this M-file are shown in Listing 8.11.

Listing 8.11: Inverted pendulum linear model and controller gains

```
A =
           0    1.0000         0         0
           0         0         0   -0.4900
           0         0         0   20.5800
           0         0    1.0000         0
B =
           0
      0.5000
     -1.0000
           0
```

```
C =

     1.0000         0         0         0
D =
         0
ans =
         5
KK =
   -56.1224   -36.7868   -35.3934  -157.6412    51.0204
```

Finally, we test the controller using the M-file shown in Listing 8.12. This M-file runs the simulation, then plots the x and θ trajectories. The plots are shown in Figure 8.11.

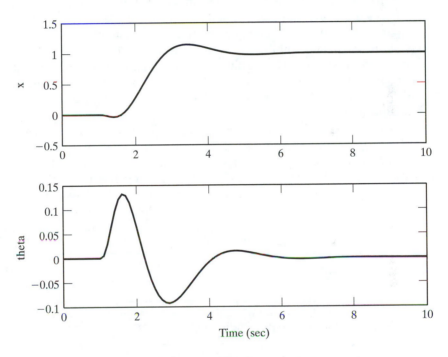

FIGURE 8.11: Inverted pendulum simulation results

Listing 8.12: M-file to test the inverted pendulum controller

```
% Run the inverted pendulum simulation
opts = simset('InitialState',[0,0,0,0,0],'Solver','ode45') ;
[t,x]= sim('InvertPendCntrl',[0:0.1:10],opts) ;
% Plot the position and pendulum angle
subplot(2,1,1);
plot(t,x(:,1));
```

```
set(gca,'FontSize',14) ;
ylabel('x');
subplot(2,1,2);
plot(t,x(:,4));
set(gca,'FontSize',14) ;
xlabel('Time (sec)') ;
ylabel('theta');
```

8.6 TRIM TOOLS

It is frequently useful to find equilibrium points of systems. As discussed in Example 8.6, an equilibrium is a point at which all state derivatives are zero. For example, a valuable technique in nonlinear systems analysis is to linearize a system about an equilibrium in order to assess the system stability in the vicinity of the equilibrium using linear systems approximations. Another situation in which equilibrium location is useful is in the assessment of the accuracy of a control system, as the steady-state behavior of the system can be thought of as an equilibrium.

The Simulink `trim` command locates equilibria and also locates partial equilibria, where we define a partial equilibrium to be a point at which selected state derivatives are zero, but other state derivatives are allowed to be nonzero. `trim` locates an equilibrium by numerically searching for a point (values of x, u, y) such that the maximum absolute value of the state derivative is minimized. If elements of the state, input, or output are fixed, the fixed elements are treated as constraints the algorithm will attempt to satisfy. In this case, the algorithm minimizes the state derivative and the constraint error. The syntax of the trim command is

$$[x,u,y,dx] = \text{trim}(\textit{model},x0,u0,y0,ix,iu,iy,dx0,idx,options,t)$$

All of the output arguments and all of the input arguments, except *model*, are optional. The arguments are defined in Table 8.5.

EXAMPLE 8.10   Locating an equilibrium

Consider the differential system

$$\dot{x}_1 = x_1^2 + x_2^2 - 4$$

$$\dot{x}_2 = 2x_1 - x_2$$

We wish to locate the equilibria of this system and assess the system stability at each equilibrium. A Simulink model of this system is shown in Figure 8.12. Shown in Figure 8.13 is a state portrait for this system. We note that there appear to be two equilibria, located near $(-1,-2)$ and $(1,2)$. Furthermore, from the state portrait, it appears that the equilibrium near $(-1,-2)$ is stable, and the equilibrium near $(1,2)$ is unstable.

TABLE 8.5: `trim` Command Arguments

Argument	Meaning
x	Value of the state vector at the equilibrium. If no output arguments are specified, trim will return x.
u	Value of the input vector at the equilibrium. u will be the empty vector if the system has no Inport blocks. Source blocks are not considered inputs to the system.
y	Value of the output vector at the equilibrium. If there are no Outport blocks, y will be the empty vector.
dx	Value of the state derivative vector at the equilibrium.
model	Name of the Simulink model, enclosed in single quotes ('), and without the file extension. For example, if the model is stored in the file s_examp.mdl, the first input argument is 's_examp'. This input argument is required.
x0	Initial guess for the state vector. The algorithm will begin the search for the equilibrium at this point. Some elements of the state vector may be fixed at the value specified in x0, using argument ix. x0 must be a column matrix, for example [0;1;0;0]. If x0 is specified, all elements must be present.
u0	Initial guess for the input vector. Some elements of the input vector may be fixed at the value specified in u0, using argument iu.
y0	Initial guess for the output vector. Some elements of the output vector may be fixed at the value specified in y0, using argument iy.
ix	Vector indicating elements of the state vector that are to be fixed, that is, treated as constraints. For example, if the second and fourth elements of the state vector are to be fixed at the value specified in x0, ix would be [2,4].
iu	Vector indicating elements of the input vector that are to be fixed, that is, treated as constraints. For example, if the second and fourth elements of u0 are to be treated as constraints, iu would be [2,4].
iy	Vector indicating elements of the output vector that are to be fixed, that is, treated as constraints. For example, if the second and fourth elements of y0 are to be treated as constraints, iy would be [2,4].
dx0	State derivative vector at a partial equilibrium point. dx0 is used in conjunction with idx to fix certain elements of the state derivative.
idx	Vector indicating elements of the state derivative vector that are to be fixed at the value specified in dx0. For example, if the third element of the state derivative is to be fixed at dx0(3), idx will be [3]. The remaining elements of the state derivative vector will be free.
options	Vector of optimization options that will be passed to the constrained optimization function. For details on options, refer to the MATLAB help screen for FOPTIONS. This argument is usually omitted or left empty ([ ]). If trim fails to converge, changing optimization parameters may help.
t	Value of time at which to locate the equilibrium. This input argument needs to be specified only if the state derivative is time dependent.

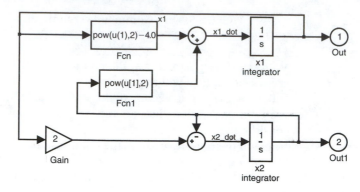

FIGURE 8.12: Simulink model of nonlinear system

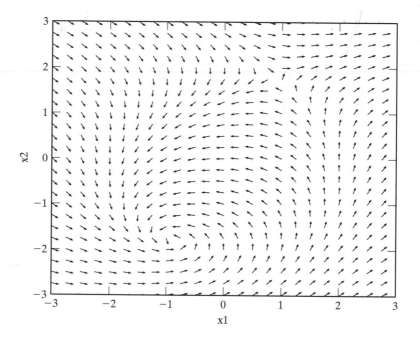

FIGURE 8.13: State portrait

To assess the stability of the equilibrium, we first locate the equilibrium points using `trim`. Next, using `linmod`, we linearize the system about each equilibrium. Finally, we use MATLAB function `eig` to compute the eigenvalues of the linearized system at each equilibrium point. An M-file that performs these operations is shown in Listing 8.13, followed by the results of executing this M-file in Listing 8.14.

Listing 8.13: M-file to assess system stability using linearization at the equilibria

```
%Locate the two equilibria for the system
%TrimXmp. Linearize the system about each equilibrium,
%then compute the eigenvalues at each equilibrium.
xa = trim('TrimXmp',[-1;-2]) ;
% Locate the equilibrium
[A,B,C,D] = linmod('TrimXmp',xa) ;
eigv_a = eig(A) ;
fprintf('Eigenvalues at (%f,%f):\n',xa) ;
eigv_a
xb = trim('TrimXmp',[1;2]) ;
% Locate the other equilibrium
[A,B,C,D] = linmod('TrimXmp',xb) ;
eigv_b = eig(A) ;
fprintf('Eigenvalues at (%f,%f):\n',xb) ;
eigv_b
```

Listing 8.14: Equilibrium assessment results

```
Eigenvalues at (-0.894427,-1.788854):
eigv_a =
  -1.3944 + 2.6457i
  -1.3944 - 2.6457i
Eigenvalues at (0.894427,1.788854):
eigv_b =
    3.4110
  -2.6222
```

Examining the results, we can see that the equilibrium near $(-1,-2)$ is actually located at $(-0.89,-1.79)$. Because the real parts of both eigenvalues at this point are negative, we can conclude that the system is stable in the vicinity of this equilibrium (this equilibrium is called a stable focus). One of the eigenvalues at the equilibrium at $(0.89,1.79)$ is negative, and the other is positive. Thus, the system behavior in the vicinity of this point should be unstable (this type of equilibrium is called a saddle point).

EXAMPLE 8.11   Spherical pendulum model

Consider the spherical pendulum shown in Figure 8.14. The system consists of a motor driving a pendulum of length l, hinged at O. The mass (m) of the pendulum is assumed to be concentrated at the end. The rotating parts of the motor are modeled as a flywheel with moment of inertia I_m. The input to the system is the motor torque (τ). Friction and aerodynamic drag are ignored. Our task is to determine the motor angular velocity ($\dot{\phi}$) that is consistent with a constant pendulum deflection (θ) of 0.5 rad.

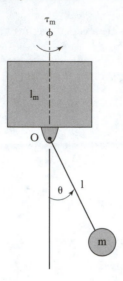

FIGURE 8.14: Spherical pendulum

We start by writing the equations of motion in terms of the shaft angle (θ) and the pendulum deflection (ϕ):

$$\ddot{\theta} = \frac{(l\,\dot{\phi}^2 \cos\theta - g)\sin\theta}{l}$$

$$\ddot{\phi} = \frac{\tau - 2\,m\,l^2\dot{\phi}\,\dot{\theta}\sin\theta\cos\theta}{I_m + m\,l^2\sin^2\theta}$$

where g represents acceleration due to gravity. In Figure 8.15, a Simulink model of this system is shown. The equations of motion are in the two function blocks.

The M-file in Listing 8.15 sets the model parameters and verifies that the ordering of the states is the same as the order of the model Outport blocks.

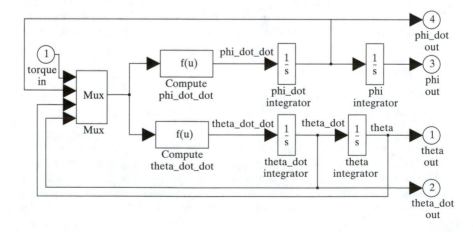

FIGURE 8.15: Spherical pendulum

Listing 8.15: Set system parameters for spherical pendulum model

```
%Set the parameters for the spherical pendulum.
%Verify the ordering of the states.
%System parameters
l = 0.3 ;               % Pendulum length (meters)
I_m = 0.75 ;            % Motor moment of inertia
                        % (kilograms*meters^2)
g = 9.8 ;               % Gravity (meters/sec^2)
m = 0.2 ;               % Pendulum bob mass (kg)
[sizes,x0,states]=spendulap([],[],[],0) ;
states
```

The pendulum deflection at the desired equilibrium is 0.5 rad, and the motor angular velocity is not zero, so we choose the initial state vector:

```
x0=[0.5;0;0;5]
```

Because the system is conservative, we expect the input to be zero at equilibrium, so set

```
u0 = 0
```

The output vector is the same as the state vector, so we set

```
y0 = [0.5;0;0;5]
```

We will allow the state and control to vary, but we require that the output pendulum deflection be 0.5 rad. Set ix and iu to the empty vector, and set

```
iy = 1
```

The value of the derivative of motor position ($\dot{\phi}$) (the third element of the state vector) will be nonzero at equilibrium, and all the other state derivatives will be zero, so set

```
dx0 = [0;0;5;0]
```

and

```
idx = [1,2,4]
```

The M-file shown in Listing 8.16 executes the trim command with these values.

Listing 8.16: Locating the spherical pendulum equilibrium

```
%Locate an equilibrium of the spherical pendulum model
x0=[0.5;0;0;5] ;
u0 = 0.0 ;
y0 = [0.5;0;0;5] ;
ix = [] ; iu = [] ;
iy = 1  ;
dx0 = [0;0;5;0] ;
idx = [1,2,4] ;
x =trim('Spendulap',x0,u0,y0,ix,iu,iy,dx0,idx)
```

resulting in the following output:

Listing 8.17: Equilibrium location output

```
x =
     0.5000
    -0.0000
     0.0000
     6.1011
```

The last element of x is the motor angular velocity at equilibrium, computed to be 6.1011 rad/sec. You can easily verify that this value results in a balance between centrifugal force and gravity.

8.7 OPTIMIZATION TOOLBOX AND Simulink

A frequent task in the design of dynamical systems and control systems is the optimization of one or more parameters, often subject to constraints. The MATLAB Optimization Toolbox provides a variety of optimization functions for unconstrained and constrained problems. The Optimization Toolbox does not provide any Simulink blocks; it is used from within the MATLAB workspace to optimize one or more parameters in a Simulink model. Particularly useful in conjunction with Simulink are fminunc (unconstrained multivariable optimization), and fmincon (constrained optimization).

The syntax of these commands is

```
x=fminunc(fun,x0,options,p1,p2,...)
```

```
x=fmincon(fun,x0,a,b,...)
```

where *fun* is a MATLAB string containing the name of the objective function, and x0 is the initial guess for the parameter vector to be optimized. We will not discuss the syntax of the commands in detail here; there are many variations of these commands. For a detailed discussion, refer to the Optimization Toolbox documentation [1] or the MATLAB help system. However, it is worthwhile to discuss the construction of the objective function.

The goal of the optimization functions is to find a parameter vector (x) such that the value of the objective function is minimized. The objective function is a function M-file of the form

```
function f = fun(x)
```

or

```
function f = fun(x,p1,p2,...,pn)
```

where p1, p2,...,pn are optional parameters that are passed unchanged to *fun* by the optimization function. An objective function used with a Simulink model has the following structure:

1. Initialize Simulink parameters based on the values of the elements of x (the input argument passed to *fun* by the optimization function).

2. Run the Simulink model using `sim`, saving the state and possibly the output trajectory or input history in MATLAB variables.

3. Compute f (a scalar) using the saved variables. f is frequently a time integral of a function of the state, input, or output trajectories, but it could also be the extreme value of a trajectory component or the final value of a component.

It's important that `sim` options SrcWorkspace and DstWorkspace be set to `'current'` (see Table 8.3) so that parameter values set in *fun* will be used instead of values assigned to variables of the same names in the base MATLAB workspace. Furthermore, because these options must be set to `'current'`, all workspace variables used to configure Simulink blocks must be assigned values in the objective function. If it is necessary to use parameters set in the base MATLAB workspace or in an M-file that calls the optimization function, pass the parameters using the optional arguments (p1,p2,...,pn).

Given a suitable objective function, the optimization function may be called from the MATLAB command line or from within an M-file. It is helpful if the initial guess for the parameter vector to be optimized (x0) is as close to the optimal value as possible, as this frequently improves convergence. Additionally, note that the optimization functions locate local minima, so if the minimum is not unique, there is no guarantee that the desired minimum will be found. Furthermore, the optimization functions terminate when the error estimate falls below a threshold (that can be set), so starting with different initial guesses can result in slightly different results, even in the neighborhood of the same minimum.

■──

EXAMPLE 8.12   Optimizing a damped pendulum

To illustrate the use of the Optimization Toolbox, consider the damped pendulum illustrated in Figure 8.16. Suppose we wish to determine the value of the damping coefficient (c) that will minimize the objective function

$$f = \int_0^{50} (\theta^2 + \dot{\theta}^2)\, dt$$

for an initial pendulum deflection of $\frac{\pi}{3}$ radians, where $m = 1$ kg and $l = 1$ m. The equation of motion of the pendulum can be written

$$\ddot{\theta} + \frac{c}{m\,l^2}\dot{\theta} + \frac{g}{l}\sin\theta = 0$$

A Simulink model of this system is shown in Figure 8.17. A function M-file that computes the value of the objective function given c is shown in Listing 8.18.

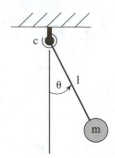

FIGURE 8.16: Damped pendulum

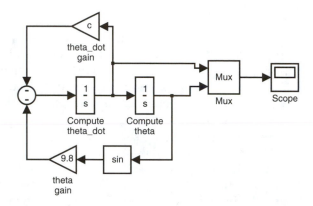

FIGURE 8.17: Simulink model of damped pendulum

Listing 8.18: Function M-file to compute damped pendulum objective function

```
function f = pdn_prf(c)
%System parameters used by \smlk model
t0 = 0 ;    % Start time
tf = 50 ;   % End time
h = .1 ;    % Time step size
opts=simset('SrcWorkspace','current') ;
% Set to use current workspace
opts=simset(opts,'DstWorkspace','current') ;
[t,x] = sim('DampedPend',[t0:h:tf],opts) ;
% Get the trajectory
[nt,nx] = size(x) ;
jp = zeros(1,nt) ;
% Preallocate performance integrand
for i = 1:nt
  jp(i) = x(i,:)*x(i,:)' ; % Integrand of objective function
end
f = h*(sum(jp)-(jp(1)+jp(nt))/2) ; % Trapezoidal integration
```

In Listing 8.19, we use `fminu` to find the optimal value of c, using a starting guess of 5.

Listing 8.19: Finding the optimal damping coefficient using fminu

```
>>  c = fminunc('pdn_prf',5)
c =
    9.4810
```

∎

8.8   OTHER TOOLBOXES USEFUL WITH Simulink

The MathWorks and others offer a number of MATLAB toolboxes that are useful in conjunction with Simulink. Some of these are analysis tools, while others provide additional Simulink blocks. The toolboxes that primarily support analysis and design include the Control System Toolbox, the System Identification Toolbox, the Frequency Domain System Identification Toolbox, and the Linear Matrix Inequality Control Toolbox. The Fuzzy Logic Toolbox and the Neural Network Toolbox provide tools for the design of fuzzy logic and neural network controllers and also provide Simulink blocks that implement the controllers. The Digital Signal Processing Blockset, Fixed Point Blockset, and Nonlinear Control Design Blockset all supply additional blocks that extend the capabilities of Simulink.

The list of products that supplement MATLAB and Simulink continues to grow. A good source of information concerning products offered by The MathWorks and others is the MathWorks Web page at www.mathworks.com.

8.9   SUMMARY

In this chapter, we discussed the use of Simulink analysis tools. First, we showed how to identify the structure of a model's state vector. Next, we showed how to run Simulink models from the MATLAB workspace using sim. Then, we used the linearization tools to find linear models of nonlinear systems and used the linear models for control system design. We also discussed using trim to locate equilibrium points. Finally, we discussed the Optimization Toolbox, and showed how to use it with a Simulink model.

REFERENCES

[1] *Optimization Toolbox User's Guide*, Version 2 (Natick, Mass.: The MathWorks, Inc., 2000).

[2] Khalil, Hassan K., *Nonlinear Systems*, 2nd ed. (Upper Saddle River, N.J.: Prentice Hall, 1996). An excellent text on the analysis and design of nonlinear systems.

[3] Ogata, Katsuhiko, *Modern Control Engineering* (Englewood Cliffs, N.J.: Prentice Hall, 1990). This book presents a thorough coverage of the standard techniques for the analysis and design of controls for continuous systems.

[4] Ogata, Katsuhiko, *Designing Linear Control Systems with MATLAB* (Englewood Cliffs, N.J.: Prentice Hall, 1993); 50–67. This book presents brief tutorials and MATLAB implementations of several important linear systems design techniques, including pole placement, state observers, and linear quadratic regulators.

[5] Shahian, Bahram, and Hassul, Michael, *Control System Design Using MATLAB* (Englewood Cliffs, N.J.: Prentice Hall, 1993). This book provides an introduction to MATLAB programming and uses MATLAB to solve many of the standard problems in classical control and modern control theory.

[6] Scheinerman, Edward C., *Invitation to Dynamical Systems* (Upper Saddle River, N.J.: Prentice Hall, 1996), 134. This text provides a good introduction to the analysis of nonlinear systems.

[7] Vidyasagar, M., *Nonlinear Systems Analysis*, 2nd ed. (Englewood Cliffs, N.J.: Prentice Hall, 1993). This book presents detailed coverage of the analysis of nonlinear systems.

9

Callbacks

In this chapter, we'll show how you can use callbacks with Simulink models. Callbacks are MATLAB commands that are automatically executed when certain events occur. Using callbacks, you can build graphical user interfaces to Simulink models. You can also use callbacks to add graphical animations.

9.1 INTRODUCTION

A *callback* is a MATLAB command that is executed when a certain event, such as opening a model or double-clicking a block, occurs. For example, normally, when you double-click a block, the block's dialog box is displayed. But consider the Slider Gain block. Double-clicking it invokes a callback that displays the slider control. In this chapter, we will discuss writing and installing callback functions.

Callbacks are closely related to MATLAB Handle Graphics. For example, when you create a menu using Handle Graphics, each menu choice generally is associated with a callback—a MATLAB command that is automatically executed when the menu choice is selected. The callback can be a simple MATLAB statement. So, if a menu choice were **Close Figure**, a suitable callback might be the MATLAB command `close(gcf)`. More often, however, callbacks are associated with M-files you write to accomplish a task corresponding to a menu choice. Because the use of callbacks is so closely related to MATLAB Handle Graphics, to get the most out of callback functions and Simulink, you'll need a solid understanding of MATLAB programming and Handle Graphics, both of which are beyond the scope of this book. In addition to the MATLAB documentation from The MathWorks [1], we recommend the text by Hanselman and Littlefield [2].

9.2 CALLBACK FUNCTION OVERVIEW

Simulink callbacks can be associated with a model or with a particular block within a model. Listed in Table 9.1 are the callbacks that can be associated with a model, and listed in Table 9.2 are the callbacks that can be associated with a particular block.

9.2.1 Callback Installation Dialog Boxes

Callbacks can be installed and associated with several standard common events using Simulink menus and callback dialogs. For example, the **File:Model Properties** page **Callbacks** provides access to a number of the model-level callback events.

TABLE 9.1: Model Callback Parameters

Parameter Name	When Executed
CloseFcn	Before model is closed. Use this callback to do any needed housecleaning when a model is closed. For example, if the model uses a custom graphical user interface, all interface windows should be closed before closing the model.
InitFcn	Before model is executed. This callback can be used to initialize MATLAB variables used to configure model or block parameters. It is particularly useful in conjunction with graphical user interfaces.
PostLoadFcn	After model is loaded. One use of this callback is to automatically start a custom graphical user interface once a model is loaded.
PostSaveFcn	After model is saved.
PreLoadFcn	Before model is loaded. This callback can be used to initialize MATLAB variables used to configure model or block parameters.
PreSaveFcn	Before model is saved.
StartFcn	Before simulation starts. This callback is executed after MATLAB workspace variables are read.
StopFcn	After the simulation stops. This callback is executed after output is written to the MATLAB workspace or files. One use of StopFcn is to automatically produce a plot when the simulation stops.

Similarly, choosing **Edit:Block Properties** and then selecting the **Callbacks** page opens a dialog box that provides access to many block callback events.

9.2.2 Installing Callbacks Using set_param

A more general method with which to install callbacks is to use the MATLAB command set_param. The syntax is

```
set_param(object, parameter, value)
```

object is a MATLAB string containing the name of the model or the path to a block. If the callback is associated with model behavior (Table 9.1), *object* is the name of the model. For example, if the model is stored in file car_mod.mdl, *object* should contain the string 'car_mod'. If the callback is associated with a block, the Simulink path to the block is used. For example, if the block of interest is named Gain_1, in subsystem Controller of car_mod, *object* should contain the string 'car_mod/Controller/Gain_1'.

parameter is a MATLAB string containing the callback parameter (the name given to the event of interest) from Table 9.1 or Table 9.2.

value is a MATLAB string containing the callback. So, for example, if the callback is a function M-file stored as set_gain.m, value would contain the string 'set_gain'.

TABLE 9.2: Block Callback Parameters

Parameter Name	When Executed
CloseFcn	When the block is closed using the close_system command. This callback is executed only when the close_system command is issued for the particular block, not for the model as a whole. Additionally, the callback will be executed whenever close_system is issued for a particular block, whether the block's dialog box is open or not.
CopyFcn	After a block is copied. This callback is recursive for subsystems. Thus, if the callback was defined for a block in a subsystem, and the subsystem is copied, the callback is executed.
DeleteFcn	Before a block is deleted. Use this callback to close any open user interface windows associated with the block. This callback is recursive for subsystems.
DestroyFcn	Before a block is destroyed. For example, when the model building command replace_block is executed, the replaced blocks are destroyed.
InitFcn	Before the block diagram is compiled and before block parameters are evaluated. This callback could be used to obtain data to set block parameters.
LoadFcn	After model is loaded. This callback is recursive for subsystems.
ModelCloseFcn	Before model is closed. This callback is recursive for subsystems. Use this callback to do any block-specific housecleaning before closing a model. For example, if there is a graphical user interface associated with the block, use this callback to close it. This is preferable to closing the interface using the model CloseFcn callback, because it eliminates the need to search for multiple instances of a block-specific interface and the need to reconfigure the model callback when a block is added or renamed.
MoveFcn	When the block is moved or resized. This callback could be used to configure a block icon based on the size or position of a block. Note that this callback is not invoked when the block is rotated or flipped.
NameChangeFcn	After the name of a block changes. Here, name means the full path name of the block. If a subsystem name is changed, all blocks in the subsystem are affected. This callback can be used in conjunction with a graphical user interface associated with a block to prevent the graphical user interface from being orphaned when the block name changes.

TABLE 9.2: Block Callback Parameters (Cont)

Parameter Name	When Executed
OpenFcn	When the block is opened. This callback is executed when you double-click a block, and it replaces the default behavior of opening the block's dialog box or opening a subsystem window.
ParentCloseFcn	Before closing a subsystem containing the block. This callback is also executed when a subsystem containing the block is created using **Edit:Create Subsystem**. This callback is the subsystem analog to the ModelCloseFcn callback. It is good practice to define this callback to behave the same as ModelCloseFcn.
PreSaveFcn	Before the model is saved. This callback is recursive for subsystems.
PostSaveFcn	After the model is saved. This callback is recursive for subsystems.
StartFcn	After the model is compiled and before the simulation starts.
StopFcn	When the simulation stops for any reason.
UndoDelete	When a block deletion is undone. If there is a CopyFcn, there should probably also be an UndoDelete.

EXAMPLE 9.1   Installing a simple callback

Consider the Simulink model shown in Figure 9.1, which we've stored as SmplClbk.mdl. The value of the Constant block is set to In_val. We wish for MATLAB to prompt the user for the value of In_val when the model is opened. A suitable callback is the MATLAB statement

```
n_val = input('Enter the value: ') ;
```

which we saved in M-file initm_1.m. To install the callback so that it is executed when the model is opened, open the model, then enter the following command at the MATLAB prompt:

```
set_param('SmplClbk','PreLoadFcn','initm_1')
```

Save the model, then close it. The next time the model is opened, the prompt

```
>> Enter the value:
```

is displayed. The value entered is assigned to In_val.

An alternative method to use to install the callback is to open the model, then choose **File:Model Properties** page **Callbacks** and enter initm_1 in field **Model pre-load function**, as shown in Figure 9.2. Again, save the model and then close it.

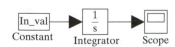

FIGURE 9.1: Simple model initialized using a callback

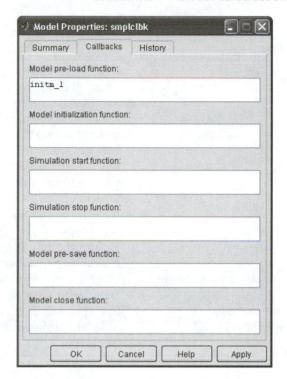

FIGURE 9.2: Simple model initialized using a callback

If we want the user to be prompted for the value of In_val at the start of each simulation instead of when the model is opened, we could install the callback using the command:

```
set_param('SmplClbk','InitFcn','initm_1')
```

Alternatively, we could enter initm_1 in **File:Model Properties** page **Callbacks** field **Simulation start function**.

9.3   MODEL CONSTRUCTION COMMANDS

Simulink provides several MATLAB commands with which to create and edit Simulink models. These commands provide the capability to create and save a new model, to add and delete blocks, to add and delete signal lines, and to get and set model and block parameters. Several of these commands useful in conjunction with callbacks are listed in Table 9.3 and are discussed below.

9.3.1   Finding the Name of the Current Block

Use gcb to find the Simulink path to the current block. For example,

```
current_block = gcb ;
```

TABLE 9.3: Simulink Model-Building Comments Useful with Callbacks

Command	Description
bdroot(*object*)	Return the name of the Simulink model associated with object. *object* is a MATLAB string containing the full Simulink path name of a block or subsystem. bdroot is used with callbacks to get the name of the model with which the current block (the block associated with the currently executing callback) is associated.
gcb	Return the full Simulink path to the current block.
gcs	Return the full Simulink path to the current system or subsystem.
get_param(*obj*, *param*)	Get the current value of a parameter associated with a system, subsystem, or block.
set_param(*obj*, *param*,*value*)	Set the value of a parameter associated with a particular system, subsystem, or block.

This function is useful when you create a graphical user interface associated with a Simulink block, as it provides a convenient and positive means to associate a Handle Graphics figure and a Simulink block. The MATLAB code fragment in Listing 9.1 will get the path name to the current block, create a figure, and store the name of the current block in the new figure's UserData

Listing 9.1: Using gcb

```
current_block = gcb ;
h_fig = figure('Position', [left bottom width height]) ;
set(h_fig, 'UserData', current_block) ;
```

9.3.2 Finding the Name of the Current Model

Use bdroot to get the name of the current model, as follows:

```
current_model = bdroot(gcb) ;
```

9.3.3 Finding the Name of the Current System

Use gcs to find the name of the current system or subsystem. For example,

```
cur_sys = gcs ;
```

Note that if a block is at the top level of a model, the commands gcs and bdroot(gcb) produce the same result. If the block is in a subsystem, the two commands produce different results.

9.3.4 Setting Parameter Values

The `set_param` command, described earlier in the context of defining callbacks, can be used to set a large number of model and block parameters. A complete list of these parameters is presented in Appendix B. More than one parameter for a particular object (system, subsystem, block) can be set using one `set_param` command. For example, the following command will set the `Tag` parameter to the string `Block A` and `Gain` parameter to 10 for a gain block:

```
set_param(gcb,'Tag','Block A','Gain','10') ;
```

Note that parameter Gain is set to the character string `'10'` and not the number 10. To set parameter `Gain` to the vector `[10 20 30]`, you could use the following command:

```
set_param(gcb,'Gain','[10 20 30]') ;
```

Command `set_param` can be executed while a simulation is in progress. Thus, if a graphical user interface changes the value of gain for a Gain block, the gain is changed immediately. This capability permits you to build interactive simulations.

SimulationCommand. A particularly useful parameter is `Simulation-Command`. This parameter may be set to the following values: `start`, `stop`, `pause`, `continue`, and `update`. Setting the value of `SimulationCommand` has the same effect as selecting the corresponding choice from the model window menu bar. For example, suppose model `SmplClbk.mdl` is currently open. Entering the command

```
set_param('SmplClbk','SimulationCommand','start')
```

at the MATLAB prompt is equivalent to choosing **Simulation:Start** from the model window menu bar.

9.3.5 Reading Parameter Values

The command `get_param` is similar to `set_param`. However, `get_param` can return only the value of one parameter at a time. Note that when `get_param` is used to read numeric data, the parameter is returned as a string. Thus, if immediately after entering the `set_param` command above, we enter the command

```
g = get_param(gcb,'Gain')
```

the following result will be displayed:

```
g =

[10 20 30]
```

but g is a character string, not a matrix. To get the values of the elements of g, use the command

```
gv = str2num(g)
```

which will result in

```
gv =

10      20      30
```

SimulationStatus. In callback functions that control a simulation, param-eter SimulationStatus can be very useful. This parameter will be set to one of the following: stopped, initializing, running, paused, terminating, or external (only if used with Real-Time Workshop). For example, if model SmplClbk.mdl is open but not running, entering, at the MATLAB prompt, the command

```
get_param('SmplClbk','SimulationStatus')
```

will result in the response

```
ans =

stopped
```

9.4 GRAPHICAL USER INTERFACES WITH CALLBACKS

Using Simulink callbacks, it is easy to build graphical interfaces to Simulink models. An excellent example of this capability is the Slider Gain block in the Math Operations block library. This block is a simple Gain block with a callback that produces the slider interface (Figure 9.3). In this section we will describe the process for building a graphical user interface using Simulink callbacks and discuss the relevant programming issues.

9.4.1 Graphic User Interface Callback

A MATLAB Handle Graphics user interface can be built using a single M-file that installs the callbacks and responds to each callback generated by the Simulink model and by the user interface. The user interface can be installed such that it is opened by double-clicking an associated block in the Simulink model. A user interface opened by double-clicking a block should respond to the following events:

1. The user interface is opened by double-clicking the associated Simulink block (callback OpenFcn). This callback should contain the MATLAB code to build

FIGURE 9.3: Slider Gain user interface

the interface figure and initialize it. The code should verify that there is not already an instance of the figure open and associated with the same Simulink block.

2. The block associated with the callback is deleted (callback `DeleteFcn`). Close the interface figure if it is open.

3. The block associated with the callback is renamed.

4. The model containing the block is closed (callback `ModelCloseFcn`). Close the figure if it is open.

5. The subsystem containing the block is closed (callback `ParentCloseFcn`). If you do not want to close the user interface in this situation, this event can be ignored.

6. A control is selected in the user interface window. There will be a callback associated with each control. These callbacks are installed when the user interface window is opened.

Additionally, it is good practice to include in the callback M-file statements to install the callbacks. Once this code is executed, the callbacks will become a permanent part of the model when the model is saved. Even though the installation code won't be needed again, it serves as a useful record of exactly how the callbacks were configured.

A convenient means by which to attach a graphical user interface to a Simulink model is to add an empty subsystem to the model. Install an `OpenFcn` callback for the subsystem block so that when the user double-clicks the subsystem block, the user interface window is opened. You can mask the subsystem to produce a custom block icon and also to provide local storage, as will be discussed shortly.

A template for a suitable M-file is shown in Listing 9.2.

Listing 9.2: Callback function M-file template `CallbackTemplate.m`

```
function CallbackTemplate(varargin)
%Callback function template
%Install this callback by invoking it with the command
%CallbackTemplate('init_block')
%at the MATLAB prompt with the appropriate model file
%open and selected.
%
%To use the template, save a copy under a new name. Then replace
%CallbackTemplate with the new name everywhere it appears.
action = varargin{1}  ;
switch action,
  case 'init_block',
    init_fcn ;                          %Block initialization function,
                                        %located in this M-file

  case 'create_fig',
    if(isempty(findobj('UserData',gcb))) %Don't open two for same block
      %Here, put all commands needed to set up the figure and its
      %callbacks.
      %Current block name is stored in block UserData
```

```
            set_param(gcb,'UserData', gcb) ;
            left =    100 ; % Figure position values
            bottom = 100 ;
            width =   100 ;
            height = 100 ;
            h_fig = figure('Position',[left bottom width height], ...
                           'MenuBar','none') ;
            set(h_fig,'UserData',gcb) ; %Save name of current block in
                                        %the figure's UserData. This is
                                        %used to detect that a CallbackTemplate
                                        %fig is open for the current block,
                                        %so that only one instance of the
                                        %figure is open at a time.
        else
            disp('Only open one instance per block can be opened')
        end
    case 'close_fig',               %Close if open when model is closed.
        h_fig = findobj('UserData',gcb) ;
        if(h_fig)                        %Is figure for current block open?
            close(h_fig) ;               %If so, close it.
        end
    case 'rename_block',     %Change the name in the figure UserData.
        oldname = get_param(gcb,'UserData') ;
        set_param(gcb,'UserData', gcb) ;
        h_fig = findobj('UserData',oldname) ;
        if(h_fig)                        %Is the figure open?
            set(h_fig,'UserData',gcb) ; %If so, change the name.
        end
    case 'UserAction1',          %Place cases for various user actions
                                 %here. These callbacks should be defined
                                 %when the figure is created.
end
%*********************************************************************
%*                          init_fcn                                *
%*********************************************************************
function init_fcn()
%Configure the block callbacks
%This function should be executed once when the block is created
%to define the callbacks. After it is executed, save the model,
%and the callback definitions will be saved with the model. There
%is no need to reinstall the callbacks when the block is copied;
%they are part of the block once the model is saved.
sys = gcs ;
block = [sys,'/InitialBlockName'] ; %Replace InitialBlockName with
                                    %the name of the block when it
                                    %is created and initialized.
                                    %This does not need to be changed
                                    %when the block is copied, as the
                                    %callbacks won't be reinstalled.
set_param(block,'OpenFcn',        'CallbackTemplate create_fig',...
               'ModelCloseFcn','CallbackTemplate close_fig', ...
               'DeleteFcn',      'CallbackTemplate close_fig', ...
               'NameChangeFcn','CallbackTemplate rename_block') ;
```

Notice that each callback has the form

```
'M-file action'
```

where *M-file* is the name of the callback M-file and *action* is a case in the switch-case code block (often called a switchyard). This syntax is identical to that used with the MATLAB eval command, and in fact, eval is useful as a quick means to test the callback M-file.

9.4.2 Programming Issues

There are several programming issues to consider when writing callback M-files. These issues are discussed next.

Avoid Embedded Names.   Don't embed block or subsystem names in the callback M-file. Instead use gcb and gcs and store the names in MATLAB variables.

Current System and Block.   MATLAB commands gcb and gcs are reliable only during the execution of Simulink callbacks. These commands are not reliable during the execution of callbacks associated with user interface controls such as pushbuttons and sliders, as there is no guarantee that the block or even the system associated with a user interface control is the current block or system. For example, gcb returns the Simulink path to the block most recently clicked, even if that block is in another open Simulink model.

Most user interface callbacks must access various parameters of the corresponding Simulink model, so it is necessary to store the name of the model, system, or block in local storage that is accessible to the callback function, and that is independent of the current system or block.

Local Data Storage.   In addition to the need to store the names of the block or system associated with a callback, it is often necessary to store local data, such as parameter values. While it is possible to store this data using global variables, that's not a good idea as it adds the risk of unintended side effects. There are two good places to store the information: UserData and masked block dialog box parameters.

Every Simulink block has a UserData parameter that can store any MATLAB variable. UserData is not stored with the model, so it cannot be used to store default data. The UserData for a Simulink block is available as long as the model is open, so it is a convenient place to store data such as the current value of model parameters. Set Simulink block UserData using the MATLAB command:

```
set_param(gcb,'UserData',value) ;
```

and read block UserData using

```
value = get_param(block,'UserData') ;
```

The value can be a single scalar, a matrix, or even a cell array or structure.

Each MATLAB Handle Graphics object [figure, axis, and user interface control (slider, pushbutton, etc.)] has its own UserData. These UserData variables are set with MATLAB Handle Graphics command set and are read with get. It is convenient to use these UserData variables to store the system and block associated with an instance of a user interface. If the user interface is created by the block callback OpenFcn, save the path to the Simulink block in the figure's UserData:

```
set(gcf,'UserData',gcb) ;
```

The UserData for Handle Graphics objects is available only as long as the object exists. For example, the UserData for a figure is available as long as the figure is open. If it is closed, the UserData is destroyed. In the callback M-file template, this fact is used as a mechanism to determine whether there is already an instance of a user interface figure associated with a particular block. Because user interface windows may be opened and closed during a Simulink session, store local data in these UserData variables only if these data are to be reinitialized each time the interface is opened.

Masked block dialog box parameters are alternatives to the block's UserData for storing user interface local data. This can be useful, as variables defined in this manner are available to the callback function and to any blocks in the masked subsystem. The variable name associated with each dialog box prompt is treated as a parameter of the masked block (the subsystem) and may be set using set_param and read using get_param. As with other block parameters, the values must be entered in the dialog box fields as strings.

EXAMPLE 9.2   Setting the value of a masked block local variable

Consider the Mask Editor **Parameters** page prompt definition shown in Figure 9.4. To set the value of the variable m associated with prompt Mass, the following statements may be entered at the MATLAB prompt or in an M-file:

```
x = 10.0 ;

x_string = num2str(x) ;

set_param('examp/subsys_a','m',x_string) ;
```

To read the contents of the field, use the following statement:

```
y=str2num(get_param('examp/subsys_a','m')) ;
```

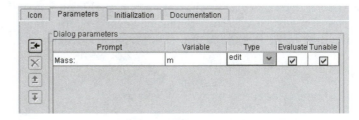

FIGURE 9.4: Mask Editor **Parameters** page prompt definition

---

EXAMPLE 9.3   Graphical user interface callback function

Consider the Simulink model shown in Figure 9.5. This system models a spring-mass-dashpot system. Double-clicking on the block labeled SetParams opens the dialog box shown in Figure 9.6. The contents of the three text fields in the dialog box are used to compute the value of gain for the two Gain blocks. The gain values are updated when the **OK** button is pressed. The parameters can be changed either before or during the execution of a simulation.

The block labeled SetParams is an empty masked subsystem. The block was created by dragging a Subsystem block from the Ports & Subsystems block library, then converting

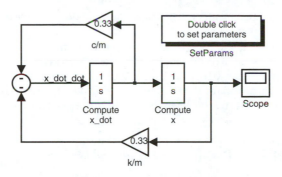

FIGURE 9.5: Simulink model with graphical user interface

FIGURE 9.6: Spring-mass-dashpot graphical user interface

Icon	Parameters	Initialization	Documentation			

Dialog parameters

Prompt	Variable	Type	Evaluate	Tunable
Mass	m	edit	☑	☑
Damping coefficient	c	edit	☑	☑
Spring constant	k	edit	☑	☑

FIGURE 9.7: Mask Editor **Parameters** page

the block into a masked subsystem using **Edit:Create Mask**. The label was created using the single command

```
disp('Double click \nto set parameters')
```

in the **Drawing commands** field on the **Icon** page of the Mask Editor. The **Parameters** page of the Mask Editor (Figure 9.7) has fields Mass, Damping coefficient, and Spring constant. The OpenFcn parameter for the masked block is set to open to the dialog box shown in Figure 9.6, so the user does not have direct access to the masked block's dialog box fields. Instead, the callback function uses these dialog box fields to store the current values of mass, damping coefficient, and spring constant. Because these values are stored in the masked block's dialog box, the parameters are saved with the model, and the most recently saved version of the parameters will be loaded in the Model Parameters dialog box fields when the model is next opened.

Listing 9.3 shows the callback function M-file that installs and responds to the callbacks. This M-file was based on the template CallbackTemplate.m. In addition to handling Simulink model events (open block, delete block, etc.), there is a callback (case 'OK_pressed') that responds to clicking the **OK** button in the Model Parameters dialog box. This callback is defined for the **OK** button when the graphical user interface is opened.

Listing 9.3: Spring-mass-dashpot callback M-file

```
function sys_set(varargin)
%********************************************************************
%*                              sys_set                            *
%********************************************************************
%Callback function for the SetParam Masked subsystem block
%This callback is installed by invoking it with the command
%sys_set('init_block')
%at the MATLAB prompt, with the appropriate model file open & selected.
%There is an additional callback associated with the sys_set figure's
%OK button, 'sys_set OK_pressed'. This callback sets the Gain blocks
%according to the current values of mass, damping, & spring constant.
action = varargin{1}  ;
switch action,
  case 'init_block',
    init_fcn ;                 %Call init_fcn, located in this M-file
  case 'create_fig',
    if(isempty(findobj('UserData',gcb)))
       create_fig ;
    end
  case 'close_fig',            %Close if open when model is closed
    h_fig = findobj('UserData',gcb) ;
    if(h_fig)                       %Is the figure for current block open?
      close(h_fig) ;                %If so, close it.
    end
  case 'rename_block',         %Change the name in the figure UserData.
    oldname = get_param(gcb,'UserData') ;
    set_param(gcb,'UserData', gcb) ;
    h_fig = findobj('UserData',oldname) ;
    if(h_fig)                       %Is the figure open?
      set(h_fig,'UserData',gcb) ; %If so, change the name.
    end
  case 'OK_pressed',                %OK button in sys_set figure
    ok_pressed ;
end


%********************************************************************
%*                              init_fcn                           *
%********************************************************************
function init_fcn()
% Configure the subsystem block callbacks
sys = gcs ; block = [sys,'/SetParams'] ;
set_param(block,'OpenFcn',       'sys_set create_fig',...
                'ModelCloseFcn','sys_set close_fig', ...
                'DeleteFcn',    'sys_set close_fig', ...
                'NameChangeFcn','sys_set rename_block') ;
```

```matlab
set_param(block,'k','1') ;       %Load initial values in the dialog
set_param(block,'m','1') ;       %box fields.
set_param(block,'c','1') ;

%***********************************************************************
%*                            create_fig                              *
%***********************************************************************
function create_fig()
%We'll get the current values of the block parameters from the
%masked block variable fields
%Current block name is stored in block UserData
set_param(gcb,'UserData', gcb) ;
mass = str2num(get_param(gcb,'m')) ;
damp = str2num(get_param(gcb,'c')) ;
spring = str2num(get_param(gcb,'k')) ;
mass_str = sprintf('%f',mass) ;        %Put values in strings
damp_str = sprintf('%f',damp) ;
spring_str = sprintf('%f',spring) ;
% Create the user interface figure and ui controls
h_fig = figure('Position',[555 406 250 169],'Tag','sys_set',...
               'MenuBar','none','NumberTitle','off', ...
               'Resize','off', ...
               'Name','Model Parameters') ;
h_mass_label = uicontrol('Parent',h_fig, ...
'Units','points', ...
'Position',[18 99.75 23.25 10], ...
'String','Mass', ...
'Style','text', ...
'Tag','mass_label');
h_damp_label = uicontrol('Parent',h_fig, ...
'Units','points', ...
'Position',[18 76.25 76.5 10], ...
'String','Damping coefficient', ...
'Style','text', 'Tag','damp_label');
h_spring_label = uicontrol('Parent',h_fig, ...
'Units','points', ...
'Position',[18 50.25 60 10], ...
'String','Spring constant', ...
'Style','text', ...
'Tag','spring_label');
h_mass_text = uicontrol('Parent',h_fig, ...
'Units','points', 'BackgroundColor',[1 1 1], ...
'Position',[97.5 99 64.5 14.25], 'String',mass_str, ...
'Style','edit', ...
'Tag','mass_string');
h_damp_text = uicontrol('Parent',h_fig, ...
'Units','points', 'BackgroundColor',[1 1 1], ...
'Position',[98.25 75 64.5 15.75], ...
'String',damp_str, ...
'Style','edit', ...
'Tag','damp_string');
h_spring_text = uicontrol('Parent',h_fig, ...
'Units','points', 'BackgroundColor',[1 1 1], ...
'Position',[98.25 49.5 65.25 12.75], ...
'String',spring_str, ...
'Style','edit', ...
'Tag','spring_string');
```

```
h_OK_button = uicontrol('Parent',h_fig, ...
'Units','points', ...
'Callback','sys_set OK_pressed', ...
'Position',[73.5 12.75 50.25 19.5], ...
'String','OK', 'Tag','OK_button', ...
'UserData',gcs );            %Save current system in OK button
set(h_fig,'UserData',gcb) ; %UserData. Save name of current block
                             %in the figure's UserData. This is
                             %used to detect that a sys_set fig
                             %is already open for the current block,
                             %so that only one instance of the figure
                             %is open at a time.


%******************************************************************
%*                         ok_pressed                           *
%******************************************************************
function ok_pressed()
% When the OK button is pressed, get the current parameter values
% and update the gain values.
h_fig = gcbf ;
h_mass = findobj(h_fig,'Tag','mass_string') ;
h_OK_button = findobj(h_fig,'Tag','OK_button') ;
mass_string = get(h_mass,'String') ;
mass = str2num(mass_string);
h_damp = findobj(h_fig,'Tag','damp_string') ;
damp_string = get(h_damp,'String') ; damp = str2num(damp_string);
h_spring = findobj(h_fig,'Tag','spring_string') ;
spring_string = get(h_spring,'String') ;
spring = str2num(spring_string);
% We could put close(h_fig) here to close the figure when OK pressed
cm_str = sprintf('%f',damp/mass) ; % Set the gain blocks
km_str = sprintf('%f',spring/mass) ;
% Note that the set_param statements that follow are looking for
% 2 particular named gain blocks. If the names of those blocks
% change, these two set_param statements must also change.
cur_sys = get(h_OK_button,'UserData') ;
cur_block = get(h_fig,'UserData') ;
set_param([get_param(cur_sys,'Name'),'/c//m'],'Gain',cm_str) ;
set_param([get_param(cur_sys,'Name'),'/k//m'],'Gain',km_str) ;
% Store values for mass, damping, spring in masked block
% dialog box fields that are not directly accessible to the user.
set_param(cur_block,'m',sprintf('%f',mass)) ;
set_param(cur_block,'c',sprintf('%f',damp)) ;
set_param(cur_block,'k',sprintf('%f',spring)) ;
```

EXAMPLE 9.4   Graphical control callback function

In this example, we will build a block that produces a graphical user interface to start and stop a simulation. Consider the Simulink model shown in Figure 9.8(a). Double-clicking on the block StartStop opens the Start/Stop control window shown in Figure 9.8(b). Whenever the simulation is not running, the label on the button is Start. Pressing the button will cause the simulation to begin. When the simulation starts, the label changes to Stop. Pressing the button when the label is Stop will cause the simulation to stop.

The block labeled StartStop is an empty masked subsystem created as described in Example 9.3. Listing 9.4 shows the M-file that installs and responds to the callbacks.

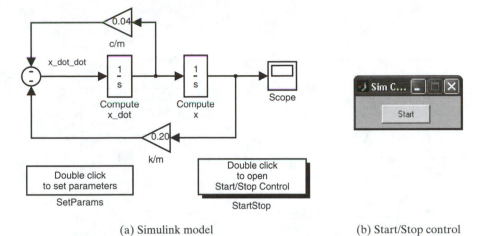

(a) Simulink model   (b) Start/Stop control

FIGURE 9.8: Model with Start/Stop control

Listing 9.4: Start/Stop control callback M-file

```
function strtstpm(varargin)
%************************************************************************
%*                        strtstpm                                     *
%************************************************************************
%Callback function for the StartStop subsystem block.
%This callback is installed by invoking it with the command
%strtstpm('init_block')
%at the MATLAB prompt, with the appropriate model file open & selected.
%There is an additional callback associated with each button.

action = varargin{1}  ;
switch action,
  case 'init_block',
    init_fcn ;                %Call init_fcn, located in this M-file.
  case 'create_fig',
    if(isempty(findobj('UserData',gcb)))
      create_fig ;
    end
  case 'close_fig',           %Close if open when model is closed.
    h_fig = findobj('UserData',gcb) ;
    if(h_fig)                  %Is the figure for current block open?
      close(h_fig) ;           %If so, close it.
    end
  case 'rename_block',        %Change the name in the figure UserData.
    oldname = get_param(gcb,'UserData') ;
    set_param(gcb,'UserData', gcb) ;
    h_fig = findobj('UserData',oldname) ;
    if(h_fig)                       %Is the figure open?
      set(h_fig,'UserData',gcb) ; %If so, change the name.
    end
  case 'Start_pressed',             %Start button
```

```
    Start_Pressed ;
  case 'Stop_pressed',              %Stop button
    Stop_Pressed ;
  case 'start_sim',                 %Something caused sim to start
    Sim_Starting ;
  case 'stop_sim',                  %Something caused sim to stop
    Sim_Stopping ;
end

%*************************************************************************
%*                          init_fcn                                    *
%*************************************************************************
function init_fcn()
% Configure the subsystem block callbacks
sys = gcs ;
block = [sys,'/StartStop'] ;
set_param(block,'OpenFcn',        'strtstpm create_fig',...
                'ModelCloseFcn','strtstpm close_fig', ...
                'DeleteFcn',      'strtstpm close_fig', ...
                'NameChangeFcn','strtstpm rename_block', ...
                'StartFcn',       'strtstpm start_sim',...
                'StopFcn',        'strtstpm stop_sim') ;

%*************************************************************************
%*                          create_fig                                  *
%*************************************************************************
function create_fig()
%Current block name is stored in block UserData.
set_param(gcb,'UserData', gcb) ;
% Create the user interface figure and ui controls.
h_fig = figure('Position',[555 406 145 44],'Tag','strtstpm',...
               'MenuBar','none','NumberTitle','off', ...
               'Resize','off', ...
               'Name','Sim Control') ;
h_Start_button = uicontrol('Parent',h_fig, ...
'Units','points', ...
'Callback','strtstpm Start_pressed', ...
'Position',[27 4.75 50.25 19.5], ...
'String','Start', ...
'Tag','Start_button', ...
'UserData',bdroot );      %Save current system in Start button UserData.
set(h_fig,'UserData',gcb) ; %Save name of current block in
                            %the figure's UserData. This is
                            %used to detect that a strtstpm fig
                            %is already open for the current block,
                            %so that only one instance of the figure
                            %is open at a time.
sim_stat = get_param(bdroot,'SimulationStatus') ;
if(strcmp(sim_stat,'running') | strcmp(sim_stat,'paused'))
  Sim_Starting ;
end

%*************************************************************************
%*                          Start_Pressed                               *
%*************************************************************************
function Start_Pressed()
%When the Start button is pressed, start the simulation.
```

```
h_fig = gcbf ;
h_Start_button = findobj(h_fig,'Tag','Start_button') ;
cur_sys = get(h_Start_button,'UserData') ;
set_param( cur_sys, 'SimulationCommand', 'start' );

%***********************************************************************
%*                        Stop_Pressed                                *
%***********************************************************************
function Stop_Pressed()
% When the Stop button is pressed, stop the simulation.
h_fig = gcbf ;
h_Start_button = findobj(h_fig,'Tag','Start_button') ;
cur_sys = get(h_Start_button,'UserData') ;
set_param( cur_sys, 'SimulationCommand', 'stop' );

%***********************************************************************
%*                        Sim_Starting                               *
%***********************************************************************
function Sim_Starting()
%Something started the sim. Make the Start_Stop button read Stop.
h_fig = findobj('UserData',gcb) ;
h_Start_button = findobj(h_fig,'Tag','Start_button') ;
set(h_Start_button,'Callback','strtstpm Stop_pressed', ...
'String','Stop') ;

%***********************************************************************
%*                        Sim_Stopping                               *
%***********************************************************************
function Sim_Stopping()
%Something stopped the sim. Make the Start_Stop button read Start.
h_fig = findobj('UserData',gcb) ;
h_Start_button = findobj(h_fig,'Tag','Start_button') ;
set(h_Start_button,'Callback','strtstpm Start_pressed', ...
'String','Start') ;
```

9.5 CALLBACK-BASED ANIMATIONS

The techniques discussed previously for associating graphical user interfaces with Simulink models can be extended to create animations. Using the techniques discussed in this section, you can build custom Simulink blocks that produce graphical output devices such as meters, or that use a graphical representation of a physical system to show the state of a simulation. An animation differs from a graphical user interface in that an animation receives the data it displays from the Simulink model as the model executes, while an interface responds to user input. There are several approaches to developing Simulink animations. In addition to the technique discussed here, animations can be built using S-Functions or using the Animation Toolbox, both of which are discussed in later chapters.

A callback-based animation consists of a masked subsystem and a callback M-file. A template animation subsystem is shown in Figure 9.9. The Zero-Order hold is used to cause the graphical output to update at specified intervals. Increasing the sample time will cause the display to update less frequently and cause the simulation to execute faster. The MATLAB Function block dialog box **MATLAB Function** field contains the command

```
anmclbk('Update_Anm',cb_var,u(1))
```

anmclbk is the name of the callback function M-file. Update_Anm is the first argument to function anmclbk, and it corresponds to the variable action in the template M-file, used to select the case to update the animation figure. cb_var is the name of the first variable in the animation subsystem **Parameters** page. We use this variable to store the handle of the animation figure when the figure is opened by double-clicking the animation block. When, during the course of the simulation, function anmclbk is called by a particular instance of the animation block, this second argument will contain the handle of the corresponding animation figure. If there is no figure corresponding to this handle (the figure either was not opened or has been closed), the callback M-file does nothing and returns. If there is a figure with that handle, the code to update the figure according to the value of the third argument [u(1)] is executed. Using this mechanism, there can be multiple instances of the animation block open at the same time, even in the same Simulink model.

The callback function M-file template CallbackTemplate.m can be used to create a callback-based animation in conjunction with the animation subsystem template shown in Figure 9.9. There should be one additional case corresponding to action 'Update_Anm', containing the code to update the figure.

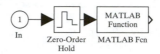

FIGURE 9.9: Animation subsystem template

EXAMPLE 9.5   Progress bar graphical animation

In this example, we will build a Simulink block that produces a progress bar that shows the progress of a simulation. The masked subsystem is shown in Figure 9.10(a), and the progress bar figure is shown in Figure 9.10(b).

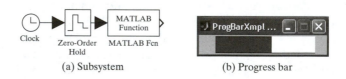

(a) Subsystem   (b) Progress bar

FIGURE 9.10: Progress bar graphical animation

Callback function M-file ProgBar.m is shown in Listing 9.5. When the user double-clicks the Progress Bar icon, function create_fig is executed. In addition to building the progress bar figure, this function obtains the current simulation start and stop times from Simulink model parameters StartTime and StopTime and saves these values in the progress bar (handle h_bar) UserData. Each time the callback function is called during the simulation by the

MATLAB Function block, the start and stop times are read from the progress bar UserData, and with the third calling argument (current simulation time), are used to update the figure.

Listing 9.5: Progress bar callback function M-file `ProgBar.m`

```
function ProgBar(varargin)
%Install this callback by invoking it with the command
%ProgBar('init_block')
%at the MATLAB prompt with the appropriate model file open & selected.
%
action = varargin{1}  ;
switch action,
  case 'init_block',
    init_fcn ;                    %Block initialization function,
                                  %located in this M-file.
  case 'create_fig',
    if(isempty(findobj('UserData',gcb)))
      create_fig ;
    else
      disp('Only open one instance per block can be opened')
    end
  case 'close_fig',             %Close if open when model is closed.
    h_fig = findobj('UserData',gcb) ;
    if(h_fig)                    %Is the figure for current block open?
      close(h_fig) ;             %If so, close it.
    end
  case 'rename_block',          %Change the name in the figure UserData.
    oldname = get_param(gcb,'UserData') ;
    set_param(gcb,'UserData', gcb) ;
    h_fig = findobj('UserData',oldname) ;
    if(h_fig)                        %Is the figure open?
      set(h_fig,'UserData',gcb) ; %If so, change the name.
    end
  case 'UpdateTime',
    if(ishandle(varargin{2}))    %Only do update if block is open.
      do_update(varargin{2},varargin{3}) ;
    end
end

%****************************************************************
%*                        init_fcn                            *
%****************************************************************
function init_fcn()
%Configure the block callbacks.
%This function should be executed once when the block is created
%to define the callbacks. After it is executed, save the model,
%and the callback definitions will be saved with the model. There
%is no need to reinstall the callbacks when the block is copied;
%they are part of the block once the model is saved.
sys = gcs ;
block = [sys,'/Progress'] ;
set_param(block,'OpenFcn', 'ProgBar create_fig',...
                'ModelCloseFcn','ProgBar close_fig', ...
                'DeleteFcn',    'ProgBar close_fig', ...
                'NameChangeFcn','ProgBar rename_block') ;
set_param(gcb,'Tag','ProgBar') ;
```

```
%************************************************************************
%*                           create_fig                               *
%************************************************************************
function create_fig()
%Current block name is stored in block UserData.
set_param(gcb,'UserData', gcb) ;
%Create the progress bar figure.
h_fig = figure('Position',[200 200 200 30], ...
               'MenuBar','none','NumberTitle','off', ...
               'Resize','off', ...
               'Name',[gcs,' Progress']) ;

set(h_fig,'UserData',gcb) ; %Save name of current block in
                            %the figure's UserData. This is
                            %used to detect that a ProgBar fig
                            %is already open for the current block,
                            %so that only one instance of the figure
                            %is open at a time.
set(h_fig,'doublebuffer','on') ;
axis([0,1,0,1]) ;
hold on ;
h_axis = gca ; set(h_axis,'Xtick',[]) ;
set(h_axis,'Ytick',[]) ;
h_bar = fill([0,0.005,0.005,0],[0,0,1,1],'r') ;
%Make the bar red.
StartTime = str2num(get_param(gcs,'StartTime')) ;
StopTime  = str2num(get_param(gcs,'StopTime'))  ;
set(h_bar,'UserData',[StartTime StopTime]) ;
%Put the progress bar handle in the masked block data field
%so that it can be passed to this function during the simulation.
set_param(gcb,'MaskValueString',num2str(h_bar,25)) ;

%************************************************************************
%*                           do_update                                *
%************************************************************************
function do_update(h_bar,current_time)
%Update the progress bar to the current simulation time.
UserData = get(h_bar,'UserData') ;
start_time = UserData(1) ;
stop_time = UserData(2)  ;
x_val = (current_time - start_time)/(stop_time - start_time) ;
set(h_bar,'Xdata',[0,x_val,x_val,0]) ;
```

9.6   SUMMARY

In this chapter, we discussed Simulink callbacks and showed how they can be used to create graphical user interfaces and to build custom animations.

REFERENCES

[1] *Using MATLAB Graphics*, Version 6 (Natick, Mass.: The MathWorks, Inc., 2000).

[2] Hanselman, Duane C., and Littlefield, Bruce R., *Mastering MATLAB 6: A Comprehensive Tutorial* (Upper Saddle River, N.J.: Prentice Hall, 2001).

10

S-Functions

In this chapter, we'll show how you can use S-Functions to build custom Simulink blocks. S-Functions allow you to incorporate existing code into a Simulink model. They can also be used in situations where it's easier to describe a subsystem algorithmically than in block diagram notation. We'll explain the development of S-Functions using MATLAB function M-files and C programming language via the MATLAB MEX-file mechanism.

10.1 INTRODUCTION

S-Functions allow you to define custom Simulink blocks using MATLAB, C, C++, Ada, or FORTRAN language code. S-Functions are useful in several situations. If there is existing code that models a portion of a system, it may be desirable to reuse that code in a Simulink model. For example, suppose you wish to try several different control system designs for a plant for which you developed a dynamics model using an M-file. You can include the dynamics model in an S-Function, then use standard Simulink blocks to model the control system. S-Functions are also useful where it is easier to describe a portion of a dynamical system algorithmically than to describe it graphically via block diagram notation. S-Functions also might improve the efficiency of a simulation, particularly in a model involving algebraic loops. (See Chapter 13 for a discussion of algebraic loops.) Finally, S-Functions provide a straightforward mechanism for adding animations to a Simulink model, a capability to be discussed in Chapter 11.

In this chapter, we will start with a general description of S-Function structure. Next we will discuss M-file S-Functions and provide examples of M-file S-Functions that have no states, that have continuous states, and that have discrete states. Then, we will discuss C S-Functions and provide C examples of the same three example subsystems. Finally, we'll briefly discuss writing S-Functions in other languages such as Ada and FORTRAN.

10.2 S-FUNCTION BLOCK

An S-Function is included in a Simulink model using the S-Function block in the User-Defined Functions block library. Shown in Figure 10.1(a) is a simple model that includes an S-Function. The block dialog box [Figure 10.1(b)] has two fields. **S-function name** contains the filename of the S-function, without an extension. This field must not be empty. S-function parameters contain any parameters required by the S-Function. If parameters are present, they must be in the form of a list, with

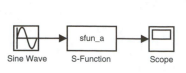

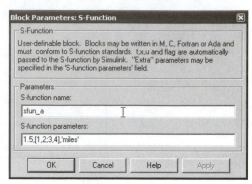

(a) Simulink model with S-Function   (b) S-Function block dialog box

FIGURE 10.1: Simple model with S-Function block

the elements separated by commas, and without brackets. For example, suppose the S-Function requires three parameters: `1.5`, a matrix `[1,2;3,4]`, and the string `'miles'`. A suitable entry in S-function parameters would be:

```
1.5,[1,2;3,4],'miles'
```

10.3 S-FUNCTION OVERVIEW

An S-Function represents a general Simulink block with input vector **u**, output vector **y**, and state vector **x**, consisting of continuous states $\mathbf{x}_c$ and discrete states $\mathbf{x}_d$. Every S-Function must include code to set the initial values of all elements of the state vector and to define the sizes of the input vector, the output vector, and the continuous and discrete components of the state vector. Also, every S-Function must compute the output

$$\mathbf{y} = \mathbf{g}(\mathbf{x}, \mathbf{u}, t, \mathbf{p}) \qquad (10.1)$$

update the discrete states (if there are any)

$$\mathbf{x}_d(k+1) = \mathbf{f}_d(\mathbf{x}, \mathbf{u}, t, \mathbf{p}) \qquad (10.2)$$

and compute the derivatives of the continuous states (if there are any)

$$\dot{\mathbf{x}}_c = \mathbf{f}_c(\mathbf{x}, \mathbf{u}, t, \mathbf{p}) \qquad (10.3)$$

Here, t is the current value of simulation time, and **p** is a vector of parameters optionally specified in S-Function block dialog box field **S-function parameters**. If an S-Function has variable sample time, it must also compute the time of the next sample hit.

The size of each of these vectors is completely arbitrary, and in particular, may be zero. So, for example, it is common for an S-Function to have inputs and outputs but no states (behavior exhibited by Gain blocks and most blocks in the Math Operations block library). An S-Function could also be used to create a custom source or sink, or a block with only continuous states (integrators) or only discrete states (delays).

In Equations (10.1) to (10.3), the functions have access to all elements of the state vector. If there are n continuous states and m discrete states, the first n elements of the state vector $\mathbf{x}$ are the current values of the continuous states $\mathbf{x}_c$, and the remaining m components of $\mathbf{x}$ are the current values of the discrete states $\mathbf{x}_d$. Simulink may call the S-Function to evaluate Equations (10.1) and (10.3) at any value of simulation time t, but Equation (10.2) is evaluated only at sample times. If there are multiple sample times, the S-Function must include logic that determines which components of $\mathbf{x}_d$ to update.

The S-Function state vector contains only the states defined by the S-Function—not all of the states in the model. Thus, if an S-Function implements a scalar integrator, its state vector will have one state. If it implements a transfer function of a third-order system, the state vector will have three states.

10.4 M-FILE S-FUNCTIONS

An M-file S-Function is a function M-file with a prescribed set of calling arguments. The first executable statement in the M-file is the function statement

```
function [sys,x0,str,ts] = sfunc_name(t,x,u,flag,p1,p2,...,pn)
```

where *sfunc_name* is the name of the S-Function. For example, if an S-Function is stored in file sfun1.m, *sfunc_name* would be sfun1 (the same as the filename, without the .m extension). The input arguments are defined in Table 10.1, and the output arguments are described in Table 10.2.

The MathWorks includes a template M-file S-Function in sfuntmpl.m, located in the MATLAB\toolbox\simulink\blocks directory in the standard Simulink installation. It is advisable to use the template as the starting point for building an S-Function, as this will eliminate some typing and allow you to start with an S-Function that follows The MathWorks conventions.

An S-Function M-file must test the value of input argument flag, and perform the corresponding operation indicated in Table 10.3. The approach taken in The MathWorks template (and here) is to use a switch-case block that calls internal

TABLE 10.1: S-Function Input Arguments

Argument	Definition
t	Current value of simulation time.
x	Current value of S-Function state vector. If there are n continuous states and m discrete states, the first n components of x are the continuous states, and the remaining m are the discrete states.
u	Current value of input vector.
flag	Flag set by Simulink each time the S-Function is called. The S-Function must test the value of flag and perform the action indicated in Table 10.3.
p1,...,pn	List of optional parameters. The number of parameters should be the same as the number of parameters in S-Function dialog box field **S-function parameters**.

TABLE 10.2: S-Function Output Argument Definitions

Argument	Definition
sys	Multipurpose output argument. The definition of sys depends on the value of flag.
x0	Initial value of S-Function state vector. x0 contains the initial value of the continuous and discrete states. x0 is needed only if flag is 0.
str	str is a placeholder output argument. It should be set to an empty matrix. str is needed only if flag is 0.
ts	Matrix of sample time, offset pairs. The matrix must have two columns. There must be at least one row. ts is needed only if flag is 0.

TABLE 10.3: S-Function flag Definition

flag Value	Action Required of the S-function
0	Initialize the sizes structure, and assign the value of the sizes structure to sys. Set x0 (initial conditions) and ts (matrix of sample times and offsets). Set str = [ ].
1	Compute the values of the derivatives of the continuous states. Set sys to the value of the derivative vector (that is, the derivative of the continuous portion of the S-Function state vector). sys is the only output argument.
2	Update the discrete states. Set sys to the new value of the discrete portion of the S-Function state vector. If there are multiple sample times, all components of the discrete state vector must be set each time flag is 2, including those that don't change. sys is the only output argument.
3	Compute outputs. Set sys to the value of the output vector. sys is the only output argument.
4	Compute the next sample time. Set sys to the value of the next sample time. sys is the only output argument.
9	Perform any necessary end-of-simulation tasks. There are no output arguments.

functions based on the value of flag. flag has six possible values. We will discuss the required behavior of the S-Function for each possible value.

10.4.1 Initialization (flag = 0)

Initialization entails four operations. The first is to set up a sizes structure and to assign this structure to the multipurpose return argument sys. Create a sizes structure using the statement

```
sizes = simsizes ;
```

TABLE 10.4: Sizes Structure

Argument	Definition
sizes.NumContStates	Number of continuous states. For example, if there are two continuous states, set sizes.NumContStates = 2 ;
sizes.NumDiscStates	Number of discrete states.
sizes.NumOutputs	Number of outputs.
sizes.NumInputs	Number of inputs.
sizes.DirFeedthrough	Set to 1 if there is direct feedthrough, otherwise 0. Direct feedthrough means that the block output [Equation (10.1)] is a function of the input. A block for which the output is an algebraic function of the input (a Gain block, trigonometric function, etc.) has direct feedthrough. A block for which the output is a function only of the states does not have direct feedthrough. Simulink uses the value of this flag to determine whether there are any algebraic loops. For more information on algebraic loops, see Chapter 13. A block that has a variable sample time has direct feedthrough.
sizes.NumSampleTimes	Number of sample time, offset pairs. Must be at least 1, even for purely continuous S-Functions.

Next, assign values to each member of sizes. Each member of sizes must be assigned a value, even if the value is zero. Listed in Table 10.4 are the members of the sizes structure. Once the sizes structure is defined, assign it to output sys using the statement

```
sys = simsizes(sizes) ;
```

The second initialization step is to set the initial condition vector x0. For example, if there are two continuous states with initial values of 1.0, and two discrete states with initial value 0.0, use the statement

```
x0 = [1.0,1.0,0.0,0.0] ;
```

The third initialization step is to set str using the statement

```
str = [ ] ;
```

The final initialization step is to create the sample time, offset matrix, ts. There must be one row in ts for each sample time, offset pair, and there must be at least one row, even for continuous S-Functions. For a continuous S-Function, use the statement

```
ts = [0,0] ;
```

As another example, suppose there are two discrete sample times. The first has a period of 0.5 s and no offset, so the sample times are at 0.0, 0.5, 1.0, ... The second has a period of 0.25 s with a 0.1 s offset, resulting in sample times of 0.1, 0.35, 0.60, . . . In this case, set

```
ts = [0.5,0 ; 0.25, 0.1] ;
```

If the sample time is to be inherited, set the sample time value to -1. The sample time will usually be inherited from the block connected to the S-Function block input. In certain situations, Simulink will detect that a longer sample time will not affect the simulation results, and the sample time (if inherited) will be set accordingly.

If the sample time is to be variable, set the sample time to -2. In this case, the S-Function will be called with `flag` = 4 to compute the time of the next sample hit.

10.4.2 Continuous State Derivatives (`flag` = 1)

If `flag` is 1, assign to `sys` the value of the derivative with respect to time of the continuous portion of the state vector. Note that if there are also discrete states, the size of `sys` will not be the same as the size of `x`, as `x` includes continuous and discrete states. If there are no continuous states, the S-Function will not be called with `flag` set to 1.

EXAMPLE 10.1   Setting continuous state derivatives

Suppose an S-Function models the nonlinear system

$$\dot{x}_1 = x_2$$

$$\dot{x}_2 = x_1 - 3x_2^2 + u_1$$

The following statements will set `sys` appropriately:

```
sys(1) = x(2);
```

```
sys(2) = x(1) - 3*x(2)^2 + u(1) ;
```

10.4.3 Discrete State Updates (`flag` = 2)

If `flag` is 2, assign to `sys` the updated value of the discrete part of the state vector. If there are multiple sample times (including hybrid systems that have a sample time of 0 for the continuous states), the S-Function must test for a sample time hit and only update those components of the discrete portion of the state vector for which the current simulation time is a sample time hit. But note that all components of the discrete portion of the state vector must be assigned a value.

EXAMPLE 10.2   Updating discrete states

First, consider the single-rate first-order discrete subsystem

$$x_1(k+1) = x_1(k) + u_1(k)$$

Use the following statement to set the new value of `sys` if `flag` is 2:

```
sys = x(1) + u(1) ;
```

Now, consider a second-order discrete subsystem with two sample times. The first discrete component is updated every 0.3 s, and the second component every 0.5 s, both with zero offset. The first state is to be updated according to the equation

$$x_1(k+1) = x_1(k) + 0.5x_2(k)$$

and the second state is to be updated according to the equation

$$x_2(k+1) = x_2(k) + u_1(k)$$

The following statements test the simulation time and update sys appropriately if the current time is a sample time hit:

```
period_1 = 0.3 ;

offset_1 = 0.0 ;

period_2 = 0.5 ;

offset_2 = 0.0 ;

sys = x ;

if abs(round(t-offset_1)/period_1-((t-offset_1)/period_1))<1.0e-8

sys(1) = sys(1) + 0.5*x(2) ;

end

if abs(round(t-offset_2)/period_2-((t-offset_2)/period_2))<1.0e-8

sys(2) = sys(2) + u(1) ;

end
```

Notice that we did not use an if..else control structure when testing for sample time hits. If we had done that, the S-Function would produce incorrect results whenever both sample times hit simultaneously. Also notice that the accuracy of this approach to locating the sample time hits is limited due to numerical precision. Greater accuracy in locating sample time hits is provided by C language S-Functions.

10.4.4 Block Outputs (flag = 3)

If flag is 3, assign to sys the value of the S-Function output [Equation (10.1)].

EXAMPLE 10.3   Setting block outputs

Suppose the output of a second order system is

$$y = x_1 + x_2$$

The following statement will set sys appropriately when flag is 3.

```
sys = x(1) + x(2) ;
```

10.4.5   Next Sample Time (`flag = 4`)

If `flag` is 4, assign to `sys` the value of the next sample time. The S-Function will be called with `flag = 4` only if the sample time is variable (set to −2).

10.4.6   Terminate (`flag = 9`)

When the simulation is complete for any reason (Stop time reached, **Simulation:Stop** selected), the S-Function is called with `flag` set to 9. The S-Function should perform any necessary end-of-simulation tasks. There is no need to assign a value to `sys`.

10.4.7   Programming Considerations

Local Data Storage.   Frequently, S-Functions require local data storage. Because an S-Function is a MATLAB function M-file, local variables must be initialized each time the S-Function is called. It is possible to preserve data between calls using global variables, but that is not a good idea. If global variables are used, there can be only one instance of a particular S-Function in a model. Otherwise, multiple instances of the S-Function would share the same storage locations. The preferred approach is to use the `UserData` parameter for the S-Function block that references the S-Function. Using the block `UserData`, there may be multiple instances of a particular S-Function in a model, because each instance of the S-Function block has its own `UserData`. `UserData` can be a scalar, a matrix, or even a cell array or structure. Therefore, there is no limit to the amount or types of data that can be stored in a block `UserData` parameter.

EXAMPLE 10.4   Saving local data in `UserData`

Suppose that an S-Function needs to store the value of time and the state vector for use the next time the S-Function is called. The following statements will save the data:

```
u_dat.time = t ;

u_dat.state = x ;

set_param(gcb,'UserData',u_dat) ;
```

The most recent values may be read from the block `UserData` using the statement

```
old_data = get_param(gcb,'UserData') ;
```

Dynamic Sizing.   Many Simulink blocks can accept inputs of varying dimension. For example, a Gain block with scalar gain can accept scalar or vector input signals, and the vector input signals can be of any dimensions. To configure an S-Function to adapt to different size input vectors, set `sizes.NumInputs` to −1. The S-Function can determine the size of the input vector when flag is 1, 2, 3, or 4 using the statement

```
size_input = size(u) ;
```

If the size of the output vector or the number of continuous or discrete states is dependent on the size of the input vector, set the appropriate member of `sizes` to -1 as well. The dimension of any of these vectors specified to be -1 is defined to be the same as the size of the input vector.

Hybrid and Multirate S-Functions.   S-Functions can be hybrid (containing both discrete and continuous states). Discrete S-Functions can have multiple sample times. There are occasions where these capabilities are useful. In general, however, it is preferable to build S-Functions that have a single purpose. This makes the S-Functions easier to develop and maintain and also makes the S-Functions more likely to be reusable for other projects.

10.4.8   M-file S-Function Examples

Three example M-file S-Functions are presented next. These examples show how to create algebraic, continuous, and discrete M-file S-Functions. Several other helpful examples can be found in the `matlab/toolbox/simulink/simdemos` directory of the standard MATLAB/Simulink installation.

EXAMPLE 10.5   M-file S-Function with no states

This example demonstrates an S-Function that has no states. The S-Function represents the algebraic equation

$$y = u_1 + u_2^2$$

The S-Function will have two inputs, one output, and no states. Because the input is passed directly to the output, the S-Function has direct feedthrough. A Simulink model that uses the S-Function is shown in Figure 10.2. S-Function block dialog box field **S-function name** contains s_xmp1. S-Function s_xmp1.m is shown in Listing 10.1:

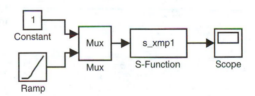

FIGURE 10.2: Simulink model with no states

Listing 10.1: M-file S-Function with no states (s_xmp1.m)

```
function [sys,x0,str,ts] = s_xmp1(t,x,u,flag)
%S-file example 1
%This is an S-file subsystem with no states. It performs
%the algebraic function y = u(1) + u(2)^2. There are two
%inputs and one output.
%
%Based on sfuntmpl.m, supplied with SIMULINK
```

```
%Copyright (c) 1990-2002 by The MathWorks, Inc.
%
switch flag,
  case 0,              % Initialization
    [sys,xO,str,ts]=mdlInitializeSizes;
  case 1,              % Compute derivatives of continuous states.
    sys=mdlDerivatives(t,x,u);
  case 2,
    sys=mdlUpdate(t,x,u);
  case 3,
    sys=mdlOutputs(t,x,u);  % Compute output vector.
  case 4,                   % Compute time of next sample.
    sys=mdlGetTimeOfNextVarHit(t,x,u);
  case 9,                   % Finished. Do any needed
    sys=mdlTerminate(t,x,u);
  otherwise                 % Invalid input
    error(['Unhandled flag = ',num2str(flag)]);
end

%*****************************************************************
%*                    mdlInitializeSizes                       *
%*****************************************************************
function [sys,xO,str,ts]=mdlInitializeSizes()
% Return the sizes of the system vectors, initial conditions, and
% the sample times and offsets.
sizes = simsizes;   % Create the sizes structure
sizes.NumContStates  = 0;
sizes.NumDiscStates  = 0;
sizes.NumOutputs     = 1;
sizes.NumInputs      = 2;
sizes.DirFeedthrough = 1;
sizes.NumSampleTimes = 1;   % At least one sample time is needed.
sys = simsizes(sizes);      % Load sys with the sizes structure.
xO  = [];    % Specify initial conditions for all states.
str = [];    % str is always an empty matrix.
ts  = [0 0]; % Initialize the array of sample times.

%*****************************************************************
%*                    mdlDerivatives                           *
%*****************************************************************
function sys=mdlDerivatives(t,x,u)
% Compute derivatives of continuous states
sys = [];   % Empty because no continuous states.

%*****************************************************************
%*                    mdlUpdate                                *
%*****************************************************************
function sys=mdlUpdate(t,x,u)
% Compute update for discrete states. If necessary, check for
% sample time hits.
sys = [];    % Empty because model has no discrete states.

%*****************************************************************
%*                    mdlOutputs                              *
%*****************************************************************
function sys=mdlOutputs(t,x,u)
% Compute output vector given current state, time, and input.
```

```
sys = [u(1) + u(2).^2];

%****************************************************************
%*                   mdlGetTimeOfNextVarHit                   *
%****************************************************************
function sys=mdlGetTimeOfNextVarHit(t,x,u)
% Return the time of the next hit for this block.  Note that
% the result is absolute time.  Note that this function is only
% used when you specify a variable discrete-time sample time.
sampleTime = [];


%****************************************************************
%*                       mdlTerminate                         *
%****************************************************************
function sys=mdlTerminate(t,x,u)
% Perform any necessary tasks at the end of the simulation.
sys = [];
```

EXAMPLE 10.6   Continuous M-file S-Function

This example illustrates using an S-Function to model a continuous system. It also illustrates the process of using a masked subsystem to supply parameters to an S-Function. Consider the inverted pendulum model from Example 8.9, shown in Figure 10.3. The equations of motion of the cart and pendulum are

$$(M + m)\ddot{x} - m\,l\,\dot{\theta}^2 \sin\theta + m\,l\,\ddot{\theta} \cos\theta = u$$

$$m\ddot{x} \cos\theta + m\,l\,\ddot{\theta} = m\,g\,\sin\theta$$

where g is the acceleration due to gravity. In Example 8.9, we manipulated the equations of motion such that $\ddot{x}$ and $\ddot{\theta}$ each appears in only one equation. An alternative approach is to rewrite the equations of motion in the form of a linear system, and solve the linear system for

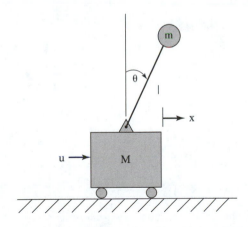

FIGURE 10.3: Cart with inverted pendulum

c and $\ddot{\theta}$. In matrix notation, the equations of motion can be written

$$\begin{bmatrix} (M+m) & m\,l\,\cos\theta \\ m\,l\,\cos\theta & m\,l \end{bmatrix} \begin{bmatrix} \ddot{x} \\ \ddot{\theta} \end{bmatrix} = \begin{bmatrix} m\,l\,\dot{\theta}^2 \sin\theta + u \\ m\,g\,\sin\theta \end{bmatrix}$$

There are three parameters in the equations of motion: the cart mass M, the pendulum mass m, and the pendulum length l. An S-Function that implements the equations of motion is shown in Listing 10.2. The function statement includes the three parameters.

The state derivatives are computed by solving the linear system for $\ddot{x}$ and $\ddot{\theta}$. Next, the value of the state vector time derivative is assigned to sys as follows:

$$\begin{bmatrix} \dot{x}_1 \\ \dot{x}_2 \\ \dot{x}_3 \\ \dot{x}_4 \end{bmatrix} = \begin{bmatrix} x_2 \\ \ddot{x} \\ x_4 \\ \ddot{\theta} \end{bmatrix}$$

The S-Function output is the state vector.

Listing 10.2: S-Function model of cart with inverted pendulum (s_xmp2.m)

```
function [sys,x0,str,ts] = s_xmp2(t,x,u,flag,M,m,l)
%S-file example 2
%This is an S-file subsystem that models a cart with
%inverted pendulum. The cart and pendulum masses and
%pendulum length are parameters that must be set in
%the block dialog box.
%
%Based on sfuntmpl.m, supplied with SIMULINK
%Copyright (c) 1990-2002 by The MathWorks, Inc.
%
switch flag,
  case 0,          % Initialization
    [sys,x0,str,ts]=mdlInitializeSizes;
  case 1,          % Compute derivatives of continuous states.
    sys=mdlDerivatives(t,x,u,M,m,l) ;
  case 2,
    sys=mdlUpdate(t,x,u);
  case 3,
    sys=mdlOutputs(t,x,u);  % Compute output vector.
  case 4,                   % Compute time of next sample.
    sys=mdlGetTimeOfNextVarHit(t,x,u);
  case 9,                   % Finished. Do any needed
    sys=mdlTerminate(t,x,u);
  otherwise                 % Invalid input
    error(['Unhandled flag = ',num2str(flag)]);
end


%***************************************************************
%*                    mdlInitializeSizes                       *
%***************************************************************
function [sys,x0,str,ts]=mdlInitializeSizes
% Return the sizes of the system vectors, initial
% conditions, and the sample times and offsets.
```

```
sizes = simsizes;    % Create the sizes structure.
sizes.NumContStates  = 4;
sizes.NumDiscStates  = 0;
sizes.NumOutputs     = 4;
sizes.NumInputs      = 1;
sizes.DirFeedthrough = 0;
sizes.NumSampleTimes = 1; % At least one sample time is needed.
sys = simsizes(sizes);    % Load sys with the sizes structure.
x0  = [0,0,0,0];    % Specify initial conditions for all states.
str = [];    % str is always an empty matrix.
ts  = [0 0]; % Initialize the array of sample times.

%***********************************************************
%*                    mdlDerivatives                       *
%***********************************************************
function sys=mdlDerivatives(t,x,u,M,m,l)
% Compute derivatives of continuous states.
g = 9.8 ;
Mass = [(M+m),m*l*cos(x(3));m*cos(x(3)),m*l] ;
x_dot_dot = Mass\[m*l*x(4)^2*sin(x(3))+u ; m*g*sin(x(3))] ;
sys = [x(2),x_dot_dot(1),x(4),x_dot_dot(2)] ;

%***********************************************************
%*                    mdlUpdate                            *
%***********************************************************
function sys=mdlUpdate(t,x,u)
% Compute update for discrete states. If necessary, check for
% sample time hits.
sys = [];    % Empty because no discrete states.

%***********************************************************
%*                    mdlOutputs                           *
%***********************************************************
function sys=mdlOutputs(t,x,u)
% Compute output vector given current state, time, and input.
sys = x ;

%***********************************************************
%*                    mdlGetTimeOfNextVarHit               *
%***********************************************************
function sys=mdlGetTimeOfNextVarHit(t,x,u)
% Return the time of the next hit for this block.  Note that
% the result is absolute time.  Note that this function is
% only used when you specify a variable discrete-time sample
% time [-2 0] in the sample time array in sampleTime = 1;
sys = [] ;

%***********************************************************
%*                    mdlTerminate                         *
%***********************************************************
function sys=mdlTerminate(t,x,u)
% Perform any necessary tasks at the end of the simulation
sys = [];
```

This S-Function can be used to build a subsystem to replace the inverted pendulum subsystem shown in Figure 8.8. To build the new subsystem, proceed as follows. Make a copy of the model in Example 8.9, and delete the Cart Model subsystem.

Drag a Subsystem block to the model window from the Ports & Subsystems block library. Open the Subsystem block, and drag Inport and Outport blocks to the subsystem window. Also drag an S-Function block to the subsystem window from the User-defined Functions block library. Connect the blocks as shown.

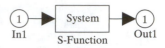

Open the S-Function block dialog box and configure the Parameters as shown. Close the Subsystem block dialog box and the subsystem window. Rename the Subsystem block `Cart Subsystem`.

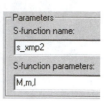

Select the Cart Subsystem block, then choose **Edit:Mask Subsystem** from the model window menu bar.

Select the **Parameters** page, and define the prompts and associated variables as shown. Note that the variables (M,m,1) are used in the S-Function block dialog box as the parameters sent to the S-Function.

Dialog parameters		
Prompt	Variable	Type
Cart mass:	M	edit ⌄
Pendulum mass:	m	edit ⌄
Pendulum length:	l	edit ⌄

Connect the output of the Sum block to the input of the new masked subsystem, and the output of the new masked subsystem to the Demux block. Configure the Gain blocks as shown in Figure 10.4. This model will produce results identical to the model in Example 8.9.

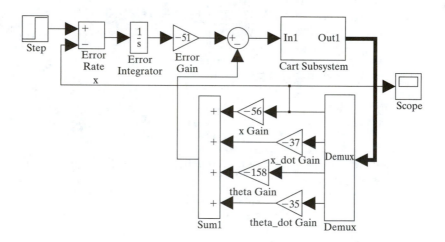

FIGURE 10.4: Inverted pendulum model using an S-Function

EXAMPLE 10.7   M-file S-Function discrete PID controller

In this example, we will use an S-Function to model a discrete PID controller. The example will also illustrate the use of local storage in an M-file S-Function. Consider the discrete PID controller used in the automobile model in Example 6.6. An S-Function that implements an equivalent controller is shown in Listing 10.3.

Listing 10.3: S-Function version of discrete PID controller (s_xmp3.m)]

```matlab
function [sys,x0,str,ts] = s_xmp3(t,x,u,flag,Kp,Ki,Kd, ...
                           L_upper,L_lower,T_samp)
% S-file example 3
% MATLAB S-file implementation of discrete PID controller.
%
%Based on sfuntmpl.m, supplied with SIMULINK
%Copyright (c) 1990-98 by The MathWorks, Inc.
%
M = 1; m = 1; l = 1;
switch flag,
  case 0,          % Initialization
    [sys,x0,str,ts]=mdlInitializeSizes(T_samp) ;
  case 1,          % Compute derivatives of continuous states.
    sys=mdlDerivatives(t,x,u) ;
  case 2,
    sys=mdlUpdate(t,x,u,L_upper,L_lower,T_samp);
  case 3,
    sys=mdlOutputs(t,x,u,Kp,Ki,Kd,T_samp);  % Compute output vector.
  case 4,                      % Compute time of next sample.
    sys=mdlGetTimeOfNextVarHit(t,x,u);
  case 9,                      % Finished. Do any needed
    sys=mdlTerminate(t,x,u);
  otherwise                    % Invalid input
    error(['Unhandled flag = ',num2str(flag)]);
end

%**********************************************************
%*                  mdlInitializeSizes                    *
%**********************************************************
function [sys,x0,str,ts]=mdlInitializeSizes(T_samp)
% Return the sizes of the system vectors, initial
% conditions, and the sample times and offsets.
sizes = simsizes;   % Create the sizes structure.
sizes.NumContStates  = 0;
sizes.NumDiscStates  = 1;
sizes.NumOutputs     = 1;
sizes.NumInputs      = 1;
sizes.DirFeedthrough = 1;
sizes.NumSampleTimes = 1;
sys = simsizes(sizes);       % Load sys with sizes structure.
x0  = [0];    % Specify initial conditions for all states.
str = [];     % str is always an empty matrix.
ts  = [T_samp 0]; % Initialize the array of sample times.
set_param(gcb,'UserData',0) ;

%**********************************************************
%*                  mdlDerivatives                        *
%**********************************************************
function sys=mdlDerivatives(t,x,u,M,m,l)
% Compute derivatives of continuous states.
sys = [];
```

```
%*************************************************************
%*                      mdlUpdate                           *
%*************************************************************
function sys=mdlUpdate(t,x,u,L_upper,L_lower,T_samp)
% Compute update for discrete states. If necessary, check
% for sample time hits.
x_new = [x(1) + T_samp*u(1)] ;
x_new = min(max(x_new,L_lower),L_upper) ; % Prevent windup.
sys = x_new ;


%*************************************************************
%*                      mdlOutputs                          *
%*************************************************************
function sys=mdlOutputs(t,x,u,Kp,Ki,Kd,T_samp)
% Compute output vector given current state, time, input.
u_prev = get_param(gcb,'UserData') ;
set_param(gcb,'UserData',u(1)) ;       % Get input at
                                       % previous sample.
u_deriv = (u(1) - u_prev)/T_samp ;     % Save current input.
sys = Kp*u(1) + Ki*x(1) + Kd*u_deriv ;


%*************************************************************
%*                  mdlGetTimeOfNextVarHit                  *
%*************************************************************
function sys=mdlGetTimeOfNextVarHit(t,x,u)
% Return the time of the next hit for this block.
sys = [] ;


%*************************************************************
%*                     mdlTerminate                         *
%*************************************************************
function sys=mdlTerminate(t,x,u)
% Perform any necessary tasks at the end of the simulation.
sys = [];
```

The discrete PID controller has one discrete state associated with the discrete integrator, and no continuous states. The S-Function requires parameters for proportional gain K_p, integral gain K_i, and derivative gain K_d. As in Example 6.6, upper L_{lower} and lower L_{upper} limits will be set on the integrator output to eliminate windup. Sample time will also be an input parameter.

Recall from Example 6.6 that the input to the controller is the difference between the desired plant output (in this case automobile speed) and the actual value. The controller output is the sum of the proportional term, the integral term, and the derivative term. For consistency with S-Function notation, we will represent the input error signal with the variable u and the controller output signal with the variable y. The integral term requires an approximation to

$$y_i = K_i \int_0^t u(\tau)d\tau$$

The state update equation will approximate the integral using forward Euler integration. Thus, the state update is

$$x(k+1) = x(k) + T\,u(k)$$

where T is the sample time. The state variable x is the current approximate value of the integral. To implement the saturation limits, $x(k+1)$ is constrained to be between L_{lower} and

TABLE 10.5: Prompt Definitions for Discrete PID Subsystem

Prompt	Variable
Proportional gain:	Kp
Integral gain:	Ki
Derivative gain:	Kd
Upper saturation limit:	L_upper
Lower saturation limit:	L_lower
Sample period:	T_samp

L_{upper}. As in Example 6.6, the derivative of the error will be approximated numerically as

$$\dot{u} \approx \frac{u(k) - u(k-1)}{T}$$

so it is necessary to store the value of the input from the previous sample. During initialization (flag = 0), the block UserData value is set to 0. Each time the output is computed (flag = 3), the previous value of the input is read from the block UserData and stored in variable u_prev. Then, the current value of the input u is stored in the block UserData parameter for use on the next sample hit.

The output of the controller is the sum of the proportional term $K_p u$, the integral term $K_i x$, and the derivative term $K_d \dot{u}$.

This S-Function is installed in the automobile model in a manner identical to that used in the previous example for the cart subsystem. The S-Function block **S-function parameters** field should contain Kp,Ki,Kd,L_upper, L_lower,T_samp. Enclose the S-Function block in a masked subsystem, and add the prompts and associated variables shown in Table 10.5 to the **Parameters** page for the masked subsystem.

10.5   C LANGUAGE S-FUNCTIONS

C language S-Functions are built using the MATLAB MEX-file mechanism. Simulink supports two C language S-Function calling conventions, Level-1 and Level-2. Level-1 S-Functions are compatible with Simulink 2, Simulink 3, Simulink 4, and Simulink 5. Level-2 S-Functions are compatible with Simulink 3, Simulink 4, Simulink 5, and only the final release of Simulink 2. Level-1 S-Functions are supported for the purposes of backwards compatibility. Level-2 S-Functions provide significantly greater capability than Level-1, provide a more consistent programming interface, and are no more difficult to code and debug. Therefore, we will only describe Level-2 S-Functions here. A similar discussion of the Level-1 conventions is presented in [1].

C language S-Functions provide all the capabilities available via M-File S-Functions, such as dynamic sizing and multiple sample times. C language S-Functions can also be written to provide capabilities not directly available within M-File S-Functions. In particular, C language S-Function blocks can have multiple input ports and multiple output ports.

C language S-Functions perform the same tasks as M-file S-Functions, but the flag mechanism is not used. Instead, a C language S-Function is a single C source file that contains a set of functions with specified names. Simulink automatically calls the functions as the simulation progresses. Certain housekeeping functions are required in all C language S-Functions. Listed in Table 10.6 are these mandatory simulation operations and the names of the C functions that are called to perform these tasks. There are also a number of optional simulation operations, listed in Table 10.7. Each optional operation is activated via a #define statement.

TABLE 10.6: Mandatory C Language S-Function Functions

Functions Called	Action Required of the S-Function
mdlInitializeSizes	Initialize the sizes portion of the simulation structure using the ssSet macros.
mdlInitializeSampleTimes	Initialize the sample times and offsets.
mdlOutputs	Compute outputs.
mdlTerminate	Perform any necessary end-of-simulation tasks.

TABLE 10.7: Optional C Language S-Function Functions

Functions Called	Action Required of the S-Function
mdlSetInputPortWidth	Set the widths of the input ports. Ordinarily, the port widths are set in mdlInitializeSizes. You can use this function when the port widths are inherited (set to DYNAMICALLY_SIZED) and you don't wish to use the default scalar expansion rules.
mdlSetOutputPortWidth	Set the widths of the output ports. Ordinarily, the port widths are set in mdlInitializeSizes. You can use this function if the port widths are inherited (set to DYNAMICALLY_SIZED) and you don't wish to use the default scalar expansion rules.
mdlSetInputPortSampleTime	Set the sample times for input ports. Ordinarily, sample times are set in mdlInitializeSampleTimes. You should use this function when sample times are inherited (set to DYNAMICALLY_SIZED).
mdlSetOutputPortSampleTime	Set output port sample times. Ordinarily, sample times are set in mdlInitializeSampleTimes. You should use this function when sample times are inherited (set to DYNAMICALLY_SIZED).
mdlSetWorkWidths	Set the widths of the work vectors. Ordinarily, these are set in mdlInitializeSizes. Use this function if work vector widths are a function of port sizes (set to DYNAMICALLY_SIZED in mdlInitializeSizes).

TABLE 10.7: Optional C Language S-Function Functions (Cont)

Functions Called	Action Required of the S-Function
mdlInitializeConditions	Set the initial value of the continuous and discrete components of the state vector. This function is needed if there are any states. The function is called again if the block is in an enabled subsystem configured to reinitialize when enabled.
mdlStart	Perform necessary operations before the simulation begins. This function is executed exactly once, even if the block is in an enabled subsystem. Generally, this function should not be used to set initial conditions, but it can be used to initialize work vectors.
mdlCheckParameters	Check block parameters. This function should be called once at the beginning of the simulation, usually in mdlInitializeSizes. Simulink will automatically call this function any time parameters change during a simulation.
mdlProcessParameters	Process block parameters. Simulink will call this function after mdlCheckParameters. This function should also be called in mdlStart to set the parameters at the beginning of a simulation.
mdlGetTimeOfNextVarHit	Compute the time of the next sample.
mdlUpdate	Update the values of the discrete components of the state vector. This will be called once for each sample hit. If there are multiple sample times, it is necessary to include logic to determine which components to update on a particular hit.
mdlDerivatives	Set the values of the derivative of the continuous components of the state vector.

The structure of a C S-function file is illustrated in Listing 10.4 and is discussed in detail in the sections that follow. Notice in Listing 10.4 that all the optional functions are shown along with the corresponding #defines. For any optional functions not needed, delete the corresponding #define and if desired, the function template as well. A template C S-Function is in the simulink/src subdirectory in file sfuntmpl.c. A thoroughly commented template is in file sfuntmpl.doc, found in the same directory.

An alternative to starting with the template is to use the S-Function Builder block from the User-Defined Functions block library. Double-clicking this block to open the S-Function Builder window makes it easy to specify many S-Function characteristics. We will illustrate C language S-function programming using the templates, as it is easier to understand the structure and operation of the code if it is presented in context.

Listing 10.4: C S-Function file structure (s_tpl_12.c)

```c
/* *********************************************************/
/*                        s_tpl_l2                       */
/* This template illustrates the structure of a          */
/* Level 2 C S-Function file                             */
/* Based on MathWorks template file                      */
/* Copyright (c) 1990-2002 by The MathWorks, Inc.        */
/* All Rights Reserved                                   */
/* *********************************************************/
#define S_FUNCTION_NAME  your_sfunction_name_here
#define S_FUNCTION_LEVEL 2
#include "simstruc.h"
#define MDL_CHECK_PARAMETERS
static void mdlCheckParameters(SimStruct *S)
{
/* Verify that the model parameters are correct. */
}
#define MDL_PROCESS_PARAMETERS
static void mdlProcessParameters(SimStruct *S)
{
/* Do anything necessary based on model parameters, both at the */
/* beginning of a simulation and when parameters change. */
}
static void mdlInitializeSizes(SimStruct *S)
{
/* Use the ssSet macros to initialize the sizes structure */
}
#define MDL_SET_INPUT_PORT_WIDTH
void mdlSetInputPortWidth(SimStruct *S, int portIndex, int width)
{
/* Verify that the width is acceptable for port portIndex */
/* If it is, set the width using ssSetInputPortWidth      */
}
#define MDL_SET_OUTPUT_PORT_WIDTH
void mdlSetOutputPortWidth(SimStruct *S, int portIndex, int width)
{
/* Verify that the width is acceptable for port portIndex  */
/* If it is, set the width using ssSetOutputPortWidth      */
}
#define MDL_SET_INPUT_PORT_SAMPLE_TIME
static void mdlSetInputPortSampleTime(SimStruct *S, int_T   portIdx,
                 real_T   sampleTime,  real_T   offsetTime)
{
/* Verify that the specified sample time and offset for the  */
/* specified port are acceptable. If they are, set them using */
/* the ssSet macros. */
}
#define MDL_SET_OUTPUT_PORT_SAMPLE_TIME
static void mdlSetOutputPortSampleTime(SimStruct *S, int_T   portIdx,
                 real_T   sampleTime,  real_T   offsetTime)
{
/* Verify that the specified sample times and offset are     */
/* acceptable. If they are, set them using the ssSet macros. */
}
static void mdlInitializeSampleTimes(SimStruct *S)
{
/* Use ssSet macros to define sample times and offsets.    */
}
#define MDL_SET_WORK_WIDTHS
#if defined(MDL_SET_WORK_WIDTHS) && defined(MATLAB_MEX_FILE)
static void mdlSetWorkWidths(SimStruct *S)
{
/* Use ssSetNum macros to set the width of work vectors that  */
/* DYNAMICALLY_SIZED                                          */
}
```

```
#define MDL_INITIALIZE_CONDITIONS
static void mdlInitializeConditions(SimStruct *S)
{
/* Initialize continuous and discrete states.              */
}
#define MDL_START
static void mdlStart(SimStruct *S)
{
/* Initialize anything that should be initialized exactly once.   */
}
#define MDL_GET_TIME_OF_NEXT_VAR_HIT
static void mdlGetTimeOfNextVarHit(SimStruct *S)
{
/* Compute the time of the next sample time hit.            */
ssSetTNext(S,timeOfNextHit>);
}
#define MDL_ZERO_CROSSINGS
static void mdlZeroCrossings(SimStruct *S)
{
/* Compute the time of the next zero crossing.             */
}
static void mdlOutputs(SimStruct *S, int_T tid)
{
/* Set the values of output signals.                       */
}
#define MDL_UPDATE
static void mdlUpdate(SimStruct *S, int_T tid)
{
/* Update the values of the changed (at current sample time) */
/* values of the state vector.                             */
}
#define MDL_DERIVATIVES
static void mdlDerivatives(SimStruct *S)
{
/* Compute the derivatives of the continuous states.       */
}
static void mdlTerminate(SimStruct *S)
{
/* Perform any end-of-simulation tasks.                    */
}
#ifdef  MATLAB_MEX_FILE
#include "simulink.c"
#else
#include "cg_sfun.h"
#endif
```

10.5.1 C File Header

The header of the C source file in Listing 10.4 declares the name of the S-Function (using a #define), declares the S-Function to be Level-2 (using a #define), and uses #include statements to supply declarations for the C math library and the S-Function data structures. The format of the #define statements is

```
#define S_FUNCTION_NAME s_fun_name
```

```
#define S_FUNCTION_LEVEL 2
```

where s_fun_name is the name of the C S-Function file and also the name of the S-Function being built. In the example in Listing 10.4, the C source file is named s_tpl_12.c.

The #define and the inclusion of simstruc.h are always required. Inclusion of math.h is usually required, but it is possible to build an S-Function that does not need the math library.

10.5.2 Initializing the Simulation Structure

Simulink will call function `mdlInitializeSizes` during simulation initialization with a pointer (`S`) to a simulation structure. A separate instance of the simulation structure is maintained for each instance of each S-Function in a model. Function `mdlInitializeSizes` initializes the size elements of the simulation structure using a set of macros defined in `simstruc.h`. The macros that must be included are defined in Table 10.8. Each macro has two or three arguments. The arguments to two-argument macros are a pointer to the simulation structure (`S`) and an integer set to the size parameter being set. For example, to set the number of continuous states to 2, use the following C code:

```
ssSetNumContStates(S, 2) ;
```

Macros that configure ports require an additional argument (the second in the argument list) to indicate which port is being specified. References to the port number use the C zero-based notation, so to reference the second port, the format of the macro would be

```
ssSetSomething(S, 1, size) ;
```

C S-Functions can be written such that various elements of the simulation structure depend on block parameters, input signal lines, or output signal lines by setting the size element to the value `DYNAMICALLY_SIZED`. At runtime, Simulink will assign storage to the input array and output array based on the size of the S-Function

TABLE 10.8: `Sizes` Initialization Macros

Macro Name	Purpose
ssSetNumContStates	Set the number of continuous states.
ssSetNumDiscStates	Set the number of discrete states.
ssSetNumOutputs	Set the number of elements in the output vector.
ssSetNumInputs	Specify the number of elements in the input vector.
ssSetDirectFeedThrough	Set to 1 if there is direct feedthrough, 0 if there is no direct feedthrough.
ssSetNumSampleTimes	Set to the number of sample times. As with M-file S-Functions, there must be at least one sample time.
ssSetNumInputArgs	Specify the number of block parameters. An M-file S-Function obtains the number of parameters from the function statement. In a C S-Function, the number of parameters must be specified.
ssSetNumIWork	Specify the number of integer work vector elements.
ssSetNumRWork	Specify the number of real work vector elements.
ssSetNumPWork	Specify the number of pointer work vector elements.

TABLE 10.9: Block Options

Option	Purpose
SS_OPTION_EXCEPTION_FREE_CODE	Can improve execution speed if no MEX functions that can cause exceptions (such as mxCalloc) are used.
SS_OPTION_RUNTIME_EXCEPTION_FREE_CODE	Use instead of the previous option if the MEX functions are used, but only in the initialization functions.
SS_OPTION_DISCRETE_VALUED_OUTPUT	Can significantly improve execution speed if the S-function produces discrete-valued outputs and is in an algebraic loop.
SS_OPTION_ALLOW_INPUT_SCALAR_EXPANSION	Explicitly declares that the S-Function supports scalar expansion.
SS_OPTION_DISALLOW_CONSTANT_SAMPLE_TIME	Prevent the S-Function from inheriting a constant sample time.
SS_OPTION_ASYNC_RATE_TRANSITION	Informs Simulink that the S-Function changes the rate of an input signal.
SS_OPTION_RATE_TRANSITION	Informs Simulink that the S-Function performs as a unit delay or zero-order hold.
SS_OPTION_NONSTANDARD_PORT_WIDTHS	If the S-Function dynamically sizes port widths, but not according to the standard Simulink rules, include this option.
SS_OPTION_PORT_SAMPLE_TIMES_ASSIGNED	Include this option if there are multiple sample times for input or output ports.

block input signal using the rules for scalar expansion. Alternatively, you can include optional functions mdlSetInputPortWidth or mdlSetOutputPortWidth and specify port widths based on input signals or block parameters. See the S-Function template sfuntmpl.doc for details.

There are a number of block options that may be set in mdlInitializeSizes to improve simulation performance or configure the S-Function block for special circumstances. The options are listed in Table 10.9. The options are set using macro ssSetOptions(S, FirstOption | SecondOption |...), where | is the bit-wise OR operator. See simstruc.h for details on each option.

EXAMPLE 10.8 **Defining multiple input ports**

Suppose an S-Function block is to have two input ports. The first is to have a scalar input signal, and the second is to accept vector inputs of any size. The block output depends on the current value of the first input but not on the current value of the second input. There is a single scalar output port. The following statements in mdlInitializeSizes will configure the ports correctly:

```
ssSetNumInputPorts(S, 2)) ;                    /* Two input ports */

ssSetInputPortWidth(S, 0, 1);                  /* First port scalar */

ssSetInputPortDirectFeedThrough(S, 0, 1);

ssSetInputPortWidth(S, 1, DYNAMICALLY_SIZED); /* Second port */

ssSetInputPortDirectFeedThrough(S, 1, 0);

ssSetNumOutputPorts(S, 1) ;                    /* One scalar output */

ssSetOutputPortWidth(S, 0, 1);
```

10.5.3 Defining Sample Times and Offsets

C S-Functions require a separate function to define sample times, in contrast to M-file S-Functions, in which all initialization tasks are performed when flag = 0. Two macros are required in function mdlInitializeSampleTimes. The first macro defines sample times, and the second defines corresponding offsets. The number of sample time, offset pairs must be the same as the number set in function mdlInitializeSizes using the ssSetNumSampleTimes macro.

Setting Sample Times.   To set a sample time, use the statement

```
ssSetSampleTime(S, n, T) ;
```

where n represents the ordinal number of the sample time being set (starting at 0), and T is the sample time. So to set the first sample time to 0.5, use the following statement:

```
ssSetSampleTime(S, 0, 0.5) ;
```

The header file simstruc.h includes #defines for three special values of sample time. Use CONTINUOUS_SAMPLE_TIME if the subsystem is continuous. Use INHERITED_SAMPLE_TIME if the sample time is to be inherited from the driving signal. Use VARIABLE_SAMPLE_TIME if the sample time is variable. Using these names will make the S-Function easier to understand.

Inherited Sample Times. If it is necessary to set sample times for input and output ports explicitly as a function of signal widths or parameters, you can use optional functions mdlSetInputPortSampleTime and mdlSetOutputPortSampleTime. In mdlInitializeSampleTimes, define the sample time to be inherited. Then add the function to set the input or output (or both) sample times for each port using the template found in sfuntmpl.doc.

Setting Sample Offsets.   To set a sample offset, use the statement

```
ssSetOffsetTime(S, n, offset) ;
```

where n represents the ordinal number of the sample time being set (starting at 0), and offset is the value of the offset. So to set the first sample time offset to 0.2, use the following statement:

```
ssSetOffsetTime(S, 0, 0.2) ;
```

10.5.4   Accessing Model Variables

It is usually necessary to read and set block state variables, inputs, and outputs. All of these are stored in the simulation structure, and macros are provided to access them.

Accessing State Variables.   Continuous and discrete state variables are stored in separate state vectors. Both are stored in special data type real_T. To obtain pointers to the continuous portion of the block state vector, the discrete portion, and the state derivative, use the following statements:

```
real_T *x_cont, *x_disc, *dx ;

x_cont = ssGetContStates(S) ;    /* Pointer to continuous states */

x_disc = ssGetDiscStates(S) ;    /* Pointer to discrete states */

dx = ssGetdX(S) ;                /* Pointer to derivative */
```

Accessing Block Inputs and Outputs.   Block inputs are accessed using a special pointer data type InputRealPtrsType. Block outputs are of type real_T. In the macro call to access inputs and outputs, there is an additional argument that specifies which input or output port is being accessed. In keeping with C conventions, the first port is Port 0, the second Port 1, and so on. The following code illustrates reading inputs and setting outputs:

```
InputRealPtrsType uPtrs ;

real_T *y, u1 ;

/* Get pointer to Port 1 signals */

uPtrs = ssGetInputPortRealSignalPtrs(S,0);

u1 = *uPtrs[0] ;              /* Read input from input Port 1 */

/* Get pointer to output Port 1 */

y = ssGetOutputPortRealSignal(S,0);

*y = u1/2.0              /* Set value of 1st output element */
```

10.5.5 Setting Initial Conditions

C S-Functions require a separate function (mdlInitializeConditions) to set initial conditions. The function should assign an initial value to each element of the state vector. For example, if there are two elements of the state vector, the first of which has initial value 0.0, and the second of which has initial value 1.0, the following statements could be used to initialize the continuous states:

```
real_T *x0 ;

x0 = ssGetContStates(S) ;        /* Pointer to continuous states */

*x0 = 0.0 ;

*(x0+1) = 1.0 ;
```

Discrete states are initialized in a similar manner, using ssGetDiscStates(S).

If the state vector is dynamically sized, use macros ssGetNumContStates(S) and ssGetNumDiscStates(S) to determine the size of each portion of the state vector. For example, suppose an S-Function is continuous, and all components of the initial condition vector are to be set to 0.0. The following statements will set x0 appropriately:

```
int nCont, i ;

real_T *x0 ;

nCont = ssGetNumContstates(S) ;

for(i = 0 ; i < nCont ; i++) *(x0+i) = 0.0 ;
```

Remember that while the state vector for M-file S-Functions consists of the continuous states followed by the discrete states, C-language S-Functions store the continuous and discrete states separately.

10.5.6 Setting the Output Vector

Function mdlOutputs sets the model output vector. If the output vector is dynamically sized, use macros ssGetNumOutputPorts(S) to find the number of output ports and ssGetOutputPortWidth(S, port) to find the size of the output vector for a particular port. Use macro ssGetT(S) to determine the current simulation time, if needed.

EXAMPLE 10.9   Dynamically sized output port

Suppose we wish to model a subsystem with dynamically sized input vector of dimension m and one continuous state variable. The state variable is defined by the differential equation

$$\dot{x} = \sum_{i=1}^{m} u_i$$

The single vector subsystem output is

$$y_i = u_i x, \qquad 1 \le i \le m$$

The following statements will set the output vector correctly:

```
int_T m, i;

InputRealPtrsType uPtrs ;

real_T *y, *x ;

x = ssGetContStates(S) ; /* Pointer to continuous states */

/* Get pointer to port 1 signals */

uPtrs = ssGetInputPortRealSignalPtrs(S,0);

/* Get pointer to output port 1 */

y = ssGetOutputPortRealSignal(S,0);

m = ssGetOutputPortWidth(S, 0) ;

for(i = 0 ; i < m ; i++) *(y + i) = *uPtrs[i] * (*x) ;
```

10.5.7　Updating the Discrete States

Function `mdlUpdate` is used to update the discrete states. If the S-Function is multirate or hybrid, use the techniques discussed in Section 10.4.3 to select which components of the discrete portion of the state vector to update. Use macro `ssGetNumDiscStates(S)` as needed to obtain the size of the block discrete state vector. To determine the current value of simulation time, use macro `ssGetT(S)`.

10.5.8　Computing the State Derivatives

Function `mdlDerivatives` must compute the derivatives of the continuous states. Use macro `ssGetNumContStates(S)` if the continuous portion of the state vector is dynamically sized.

EXAMPLE 10.10　Setting C S-Function state derivatives

The following statements will compute and set the state derivative defined in Example 10.9:

```
real_T *dx ;

int_T ninp, n ;

InputRealPtrsType uPtrs ;

dx = ssGetdX(S) ;        /* Pointer to derivative */

/* Get pointer to port 1 signals */

uPtrs = ssGetInputPortRealSignalPtrs(S,0);

*dx = 0 ;
```

```
ninp = ssGetInputPortWidth(S, 0) ;

for(n = 0 ; n < ninp ; n++) *dx = *dx + *uPtrs[n] ;
```

10.5.9   End-of-Simulation Tasks

Function `mdlTerminate` should perform any necessary end-of-simulation tasks. The only input argument to `mdlTerminate` is a pointer to the S-Function simulation structure (S).

10.5.10   C S-Function File Trailer

The C S-Function file trailer consists of the `#ifdef .. #endif` compiler directives shown in the last five lines of Listing 10.4. These statements are necessary in order for the MATLAB MEX-file mechanism to work properly.

10.5.11   Programming Considerations

The many functions in the MATLAB MEX application programming interface are accessible from within C S-Functions. Some of the MEX functions most useful in C S-Functions are discussed in this section. Refer to the API documentation [2] and the online API documentation for more detailed coverage of MEX programming.

Persistent Local Data Storage.   The S-Function mechanism provides persistent storage for real, integer, and pointer data. The persistent storage for each instance of an S-Function block in a Simulink model is unique. Therefore, there is no conflict if there are multiple instances of an S-Function block in the same model. The persistent storage variables are called work vectors. The size of each work vector (real, integer, pointer) must be set in `mdlInitializeConditions` using macros `ssSetNumRWork`, `ssSetNumIWork`, and `ssSetNumIWork`. For example, to set the number of real work vector elements to 3, use the statement

```
ssSetNumRWork(S, 3) ;
```

Values are stored in the work vectors using `ssSetRWorkValue`, `ssSetIWorkValue`, and `ssSetPWorkValue`. The syntax is

```
ssSetRWorkValue(S, R_index, value) ;
```

where `R_index` is the index into the work vector (real in this case), and `value` is the value being stored. Read data from the work vector using `ssGetRWorkValue`, `ssGetIWorkValue`, and `ssGetPWorkValue`. The syntax is

```
value = ssGetRWorkValue(S, R_index) ;
```

Table 10.10 lists `simstruc.h` macros that access work vectors.

Accessing Block Parameters.   You can access S-Function block parameters in C S-Functions using the `ssGetSFcnParam(S, arg_num)` macro. `arg_num` is the ordinal argument number, with the first argument 0, the second 1, and so on.

TABLE 10.10: Work Vector Macros

Macro Name	Purpose
ssGetIWorkValue(S, iworkIdx)	Get the value of an element of the integer work vector.
ssSetIWorkValue(S, iworkIdx,iworkValue)	Set the value of an element of the integer work vector.
ssGetRWorkValue(S, rworkIdx)	Get the value of an element of the real work vector.
ssSetRWorkValue(S, rworkIdx,rworkValue)	Set the value of an element of the real work vector.
ssGetPWorkValue(S, pworkIdx)	Get a pointer from the pointer work vector.
ssSetPWorkValue(S, pworkIdx,pworkValue)	Set a pointer in the pointer work vector.

ssGetSFcnParam returns a pointer compatible with MEX API functions mxGetPr and mxGetPi. mxGetPr returns a pointer to the real part of the parameter, and mxGetPi returns a pointer to the imaginary part. The recommended programming technique is to specify at the top of the S-Function file a #define for each parameter. So, if there are two scalar parameters, you could use the following statements in the C S-Function file header:

```
#define param_1 *mxGetPr(ssGetSFcnParam(S,0))

#define param_2 *mxGetPr(ssGetSFcnParam(S, 1))
```

Then, in the function that uses the arguments, use the following statements:

```
double p1, p2 ;

p1 = param_1 ;

p2 = param_2 ;
```

The simplest way to access the block parameters is to place the calls to the ssGetSFcnParam macro directly in the functions that use parameters. Using this simple approach, the parameters are read each time they are used, and if the parameters change during a simulation, the changes will take effect immediately. It is also possible to write an S-Function such that parameter values are stored in work vectors and read only at the beginning of a simulation and when Simulink detects that a parameter changed. To implement this additional capability, two functions are needed. Function mdlCheckParameters should contain code to read and verify parameters. mdlProcessParameters should contain code to respond to parameter changes, say, to save them in work vectors. If these parameter-handling functions are present, your S-Function is responsible for calling them once at the beginning of the simulation, via a call to mdlCheckParameters in mdlInitializeSizes and a call to mdlProcessParameters in mdlStart. After the simulation begins, Simulink will automatically call these functions whenever a parameter changes. Example 10.13 illustrates the simple approach, and Example 10.14 illustrates the advanced approach.

EXAMPLE 10.11 Accessing a work vector

Suppose we wish to create an S-Function that produces as its output the maximum value its scalar input has reached thus far during the simulation:

$$y(t) = \max(u(\tau)), \qquad 0 \le \tau \le t$$

In the function `mdlInitializeSizes`, include the statement

```
ssSetNumRWork(S, 1) ;
```

In the function `mdlInitializeConditions`, include the statement

```
ssSetRWorkValue(S, 0, -9.9E20) ;
```

to allocate space for a work vector that will hold a single real number and to initialize the contents of the work vector to a value that is certain to be less than the initial value of the input signal. Listing 10.5 illustrates the output function.

Listing 10.5: Accessing the work vector (`mdlOutputs.c`)

```c
static void mdlOutputs(SimStruct *S, int_T tid)
{
  InputRealPtrsType uPtrs ;
  real_T *y, u1 ;
  double u_max_prev ;
  u_max_prev = ssGetRWorkValue(S, 0) ;
  /* Get pointer to port 1 signals */
  uPtrs = ssGetInputPortRealSignalPtrs(S,0);
  /* Get pointer to output port 1 signal */
  y = ssGetOutputPortRealSignal(S,0);
  u1 = *uPtrs[0] ;
  *y = u_max_prev ;
  if(u1 > u_max_prev)
    {
    *y = u1 ;
    ssSetRWorkValue(S, 0, u1) ;
    }
}
```

Accessing the MATLAB Engine. One of the most powerful features of MEX files, including C S-Functions, is that they have access to the MATLAB computational engine. Details on calling MATLAB functions from within MEX functions are provided in the MATLAB API Guide [2] and the online MEX documentation (through the MATLAB Help Desk). To illustrate the concepts needed, we will discuss the procedure for defining a real matrix and using MATLAB to compute its eigenvalues.

To create a matrix, which we'll call A, three statements are required:

```
mxArray *A ;

double *A_dat

A = mxCreateDoubleMatrix(nrows, ncols, mxREAL) ;
```

Assuming that `nrows` and `ncols` were previously set to 2, A is now a real MATLAB matrix with two rows and two columns. Next, assign a pointer to the data in A using the statement

```
A_dat = mxGetPr(A) ;
```

The data in a MATLAB matrix is stored using the FORTRAN columnwise convention. Suppose the matrix A is defined as

$$\mathbf{A} = \begin{bmatrix} 1.5 & 3.5 \\ 2.3 & 4.8 \end{bmatrix}$$

The following statements will fill A correctly:

```
*A_dat = 1.5 ;

*(A_dat + 1) = 2.3 ;

*(A_dat + 2) = 3.5 ;

*(A_dat + 3) = 4.8 ;
```

To call a MATLAB function, use MEX function mexCallMATLAB. The syntax is

```
mexCallMATLAB(num_ret, ret_array, num_in, in_array, operation) ;
```

num_ret is the number of returned arguments. So, if the MATLAB function is called using a statement of the form `[ret_1, ret_2] = MATLAB_function(A)`, num_ret would be 2.

ret_array is a pointer to an array of MATLAB matrices. ret_array must be declared in the S-Function. In this example, ret_array should be declared as

```
mxArray *ret_array[1] ;
```

num_in is the number of input arguments, in this case 1.

in_array is a pointer to an array of MATLAB matrices. Here, in_array is declared

```
mxArray *in_array[1] ;
```

and set to point to A:

```
in_array[0] = A ;
```

The final argument, `operation`, is a string containing the name of the MATLAB function. Here, operation is `eig`. If the operation is a basic math operation such as addition, use the symbol for the operation, enclosed in quotes: `"+"`.

The function should end with statements freeing the storage allocated to the array. Function mxDestroyArray frees the storage:

```
mxDestroyArray(A) ;
```

TABLE 10.11: Useful `simstruc.h` Macros

Macro	Purpose
ssGetT(S)	Get current simulation time.
ssGetTStart(S)	Get the simulation start time.
ssGetTFinal(S)	Get the simulation stop time.
ssGetStepSize(S)	Return the current simulation stepsize.

Handling Errors. It is usually a good programming practice to check for errors and, when errors are found, report them and terminate the program so the user can correct them. Simulink provides a built-in mechanism for responding to errors in S-Functions. To display an error message and terminate, use the following two lines of code:

```
ssSetErrorStatus(S, "Error message") ;

return ;
```

The error message string must be in persistent storage, either a constant string or a static variable.

Useful Macros in `simstruc.h`. The S-Function header file `simstruc.h` provides a number of macros that provide access to S-Function parameters and variables. Some of the more useful macros that haven't been discussed are listed in Table 10.11. Refer to the header file for a complete list.

10.5.12 Compiling a C S-Function

Compile a C S-Function by entering the following command at the MATLAB prompt:

```
mex sfun_name.c
```

where `sfun_name.c` is the name of the C source file. The `mex` command will generate the compiler and linker commands necessary to produce the executable S-Function file. Compiler error messages will be displayed in the MATLAB window.

10.5.13 C S-Function Examples

This section provides C S-Function versions of the examples in Section 10.4.8. In each case, the Simulink model is configured the same for the C S-Function as for the M-file S-Function. The only difference is the name of the S-Function file in the S-Function block dialog box.

■
EXAMPLE 10.12 C S-Function with no states

This example demonstrates an S-Function that has no states. The S-Function represents the algebraic equation

$$y = u_1 + u_2^2$$

The C S-Function is shown in Listing 10.6. Notice that the C S-function is configured with two input ports, whereas for the M-file S-Function, it was necessary to multiplex the two inputs into a single vector input signal.

Listing 10.6: C S-Function with no states (s_xmp4L2.c)

```c
/* ******************************************************/
/*                     s_xmp4L2 TwoInputs              */
/* C S-File example that computes y = u(1) + u(2)^2    */
/* u(1) and u(2) are separate input ports              */
/* Based on MathWorks template file                    */
/* Copyright (c) 1990-2002 by The MathWorks, Inc.      */
/* All Rights Reserved                                 */
/* ******************************************************/
#define S_FUNCTION_LEVEL 2
#define S_FUNCTION_NAME s_xmp4L2
#include "simstruc.h"
#include "math.h"
/* ******************************************************/
/*                  mdlInitializeSizes                 */
/* ******************************************************/
static void mdlInitializeSizes(SimStruct *S)
{
    /* number of continuous states */
    ssSetNumContStates(    S, 0);
    /* number of discrete states */
    ssSetNumDiscStates(    S, 0);
    /* number of input ports      */
    if (!ssSetNumInputPorts(S, 2)) return;
    ssSetInputPortWidth(S, 0, 1);
    ssSetInputPortDirectFeedThrough(S, 0, 1);
    ssSetInputPortWidth(S, 1, 1);
    ssSetInputPortDirectFeedThrough(S, 1, 1);
    /* number of output ports */
    if (!ssSetNumOutputPorts(S, 1)) return;
    ssSetOutputPortWidth(S, 0, 1);
    /* number of sample times */
    ssSetNumSampleTimes(S, 1);
    /* number of input arguments */
    ssSetNumSFcnParams(S, 0);
    /* number of real work vector elements */
    ssSetNumRWork(S, 0);
    /* number of integer work vector elements*/
    ssSetNumIWork(S, 0);
    /* number of pointer work vector elements*/
    ssSetNumPWork(S, 0);
    ssSetNumModes(S, 0);
    ssSetNumNonsampledZCs(S, 0);
    ssSetOptions(S, 0);
}
/* ******************************************************/
/*               mdlInitializeSampleTimes              */
/* ******************************************************/
static void mdlInitializeSampleTimes(SimStruct *S)
{
    ssSetSampleTime(S, 0, CONTINUOUS_SAMPLE_TIME);
```

```
    ssSetOffsetTime(S, 0, 0.0);
}
/* ********************************************************/
/*                      mdlOutputs                     */
/* ********************************************************/
static void mdlOutputs(SimStruct *S, int_T tid)
{
InputRealPtrsType uPtrs1, uPtrs2 ;
real_T *y, u1, u2 ;
/* Get pointer to port 1 signals */
uPtrs1 = ssGetInputPortRealSignalPtrs(S,0);
/* Get pointer to port 2 signals */
uPtrs2 = ssGetInputPortRealSignalPtrs(S,1);
/* Get pointer to output port 1 signal */
y = ssGetOutputPortRealSignal(S,0);
u1 = *uPtrs1[0] ;
u2 = *uPtrs2[0] ;
*y = u1 + pow(u2,2.0) ;    /* y = u(1) + u(2)^2  */
}
/* ********************************************************/
/*                      mdlTerminate                   */
/* ********************************************************/
static void mdlTerminate(SimStruct *S)
{
}
#ifdef MATLAB_MEX_FILE
#include "simulink.c"
#else
#include "cg_sfun.h"
#endif
```

EXAMPLE 10.13   Continuous C S-Function

This example models the cart with inverted pendulum in a manner identical to Example 10.6. It demonstrates two important capabilities. Referring to Listing 10.7, notice the #defines in the C file header. These #defines illustrate using S-Function block dialog box parameters in a C S-Function. Second, this S-Function illustrates the procedure for using the MATLAB engine from within the S-Function.

Listing 10.7: C S-Function model of cart with inverted pendulum (s_xmp5L2.c)

```
/* ********************************************************/
/*                      s_xmp5L2                       */
/*   This is a C version of the S-file subsystem that  */
/*   models a cart with inverted pendulum. The cart    */
/*   and pendulum masses and pendulum length are       */
/*   parameters that must be set in the block dialog   */
/*   box.                                              */
/*                                                     */
/* Based on MathWorks template file                    */
```

```c
/* Copyright (c) 1990-2002 by The MathWorks, Inc.        */
/* All Rights Reserved                                   */
/* ******************************************************/
#define S_FUNCTION_LEVEL 2
#define S_FUNCTION_NAME s_xmp5L2
#include "simstruc.h"
#include "math.h"
#define M_macro *mxGetPr(ssGetSFcnParam(S,0))
#define m_macro *mxGetPr(ssGetSFcnParam(S,1))
#define l_macro *mxGetPr(ssGetSFcnParam(S,2))
/* ******************************************************/
/*                  mdlInitializeSizes                  */
/* ******************************************************/
static void mdlInitializeSizes(SimStruct *S)
{
    ssSetNumContStates(S, 4);
    ssSetNumDiscStates(S, 0);
    /* Input ports */
    if (!ssSetNumInputPorts(S, 1)) return;
    ssSetInputPortWidth(S, 0, 1);
    ssSetInputPortDirectFeedThrough(S, 0, 0);
    /* Output ports */
    if (!ssSetNumOutputPorts(S, 1)) return;
    ssSetOutputPortWidth(S, 0, 4);
    ssSetNumSampleTimes(S, 1);
    ssSetNumSFcnParams(S, 3);
    ssSetNumRWork(S, 0);
    ssSetNumIWork(S, 0);
    ssSetNumPWork(S, 0);
    ssSetNumModes(S, 0);
    ssSetNumNonsampledZCs(S, 0);
    ssSetOptions(S, 0);
}
/* ******************************************************/
/*              mdlInitializeSampleTimes                */
/* ******************************************************/
static void mdlInitializeSampleTimes(SimStruct *S)
{
    ssSetSampleTime(S, 0, CONTINUOUS_SAMPLE_TIME);
    ssSetOffsetTime(S, 0, 0.0);
}
/* ******************************************************/
/*                mdlInitializeConditions               */
/* ******************************************************/
#define MDL_INITIALIZE_CONDITIONS
static void mdlInitializeConditions(SimStruct *S)
{
    real_T *x0 ;
    int i ;
    x0 = ssGetContStates(S) ;
    for(i = 0 ; i < 4 ; i++) *(x0+i) = 0.0 ;
}
/* ******************************************************/
/*                     mdlOutputs                       */
/* ******************************************************/
static void mdlOutputs(SimStruct *S, int_T tid)
{
```

```c
   real_T *y, *x ;
   int i ;
   y = ssGetOutputPortRealSignal(S, 0) ;
   x = ssGetContStates(S) ;
   for(i = 0 ; i < 4 ; i++) *(y+i) = *(x+i) ; /* y = x */
}
/* ************************************************************/
/*                        mdlDerivatives                   */
/* ************************************************************/
#define MDL_DERIVATIVES
static void mdlDerivatives(SimStruct *S)
{
   real_T *dx, *x ;
   InputRealPtrsType uPtrs ;
   double u ;
   static double g = 9.8 ;    /* Acceleration due to gravity */
   double M, m, l ;            /* Model parameters */
   /* MATLAB matrix to contain the mass matrix */
   mxArray *Mass ;
   /* pointer to the data part of the matrix */
   double *Mass_v ;
   /* Store right hand side of dynamic equation */
   mxArray *x_dot_dot_rhs ;
   double *x_dot_dot_rhs_v ;  /* pointer to the data */
   mxArray *input_array[2], *out_array[1] ;
   double *x_dot_dot ;
   x = ssGetContStates(S) ; /* Pointer to state */
   dx = ssGetdX(S) ;        /* Pointer to derivative */
   /* Get pointer to port 1 signals */
   uPtrs = ssGetInputPortRealSignalPtrs(S,0);
   u = *uPtrs[0] ;
   M = M_macro ;  /* Obtain the values of the parameters */
   m = m_macro ;
   l = l_macro ;
   /* Create mass matrix */
   Mass = mxCreateDoubleMatrix(2, 2, mxREAL) ;
   Mass_v = mxGetPr(Mass) ;
   /* Load the values in mass matrix */
   *Mass_v = M + m ;
   *(Mass_v+1) = m*cos(*(x+2)) ;
   *(Mass_v+2) = m*l*cos(*(x+2)) ;
   *(Mass_v+3) = m*l ;
   /* Create rhs */
   x_dot_dot_rhs = mxCreateDoubleMatrix(2, 1, mxREAL) ;
   x_dot_dot_rhs_v = mxGetPr(x_dot_dot_rhs) ;
   *x_dot_dot_rhs_v = m*l*pow(*(x+3),2.0)*sin(*(x+2)) + u ;
   *(x_dot_dot_rhs_v+1) = m*g*sin(*(x+2)) ;
    /* Set up for call to MATLAB */
   input_array[0] = Mass ;
   input_array[1] = x_dot_dot_rhs ;
   /* Solve system */
   mexCallMATLAB(1, out_array, 2, input_array, "\\") ;
   x_dot_dot = mxGetPr(out_array[0]) ;
   /* Fill derivative vector */
   *dx = *(x+1) ;
   *(dx+1) = *x_dot_dot ;
   *(dx+2) = *(x+3) ;
```

```
    *(dx+3) = *(x_dot_dot+1) ;
    /* Free the memory allocated by MATLAB */
    mxDestroyArray(out_array[0]) ;
    /* Free the memory used for Mass and rhs */
    mxDestroyArray(Mass) ;
    mxDestroyArray(x_dot_dot_rhs) ;
}
/* ******************************************************* */
/*                        mdlTerminate                    */
/* ******************************************************* */
static void mdlTerminate(SimStruct *S)
{
}
#ifdef MATLAB_MEX_FILE
#include "simulink.c"
#else
#include "cg_sfun.h"
#endif
```

EXAMPLE 10.14   C S-Function discrete PID controller

This example presents a C S-Function (see Listing 10.8) that implements the discrete PID controller discussed in Example 10.7. This example uses S-Function block parameters to configure the controller. It also uses an S-Function work vector to store the value of the block input from the previous sample, for use in approximating the derivative of the input signal (the speed error).

Listing 10.8: C S-Function implementation of discrete PID controller
(s_xmp6L2.c)

```
/* ******************************************************* */
/*                        s_xmp6L2                        */
/*   This is a C version of the S-file subsystem that     */
/*   implements a discrete PID controller.                */
/*                                                         */
/* Based on MathWorks template file                       */
/* Copyright (c) 1990-2002 by The MathWorks, Inc.         */
/* All Rights Reserved                                    */
/* ******************************************************* */
#define S_FUNCTION_LEVEL 2
#define S_FUNCTION_NAME s_xmp6L2
#include "simstruc.h"
#include "math.h"
#define Kp_macro        ssGetRWorkValue(S, 1)
#define Ki_macro        ssGetRWorkValue(S, 2)
#define Kd_macro        ssGetRWorkValue(S, 3)
#define L_upper_macro   ssGetRWorkValue(S, 4)
#define L_lower_macro   ssGetRWorkValue(S, 5)
#define T_samp_macro    ssGetRWorkValue(S, 6)
```

```c
/* *********************************************************/
/*                    mdlInitializeSizes                  */
/* *********************************************************/
static void mdlInitializeSizes(SimStruct *S)
{
    ssSetNumContStates(    S, 0);
    ssSetNumDiscStates(    S, 1);
    /* Input ports     */
    if (!ssSetNumInputPorts(S, 1)) return;
    ssSetInputPortWidth(S, 0, 1);
    ssSetInputPortDirectFeedThrough(S, 0, 1);
    /* Output ports */
    if (!ssSetNumOutputPorts(S, 1)) return;
    ssSetOutputPortWidth(S, 0, 1);
    ssSetNumSampleTimes(S, 1);
    ssSetNumSFcnParams(S, 6);
    if (ssGetNumSFcnParams(S) != ssGetSFcnParamsCount(S))
      {
      return;
      }
    mdlCheckParameters(S) ;
    ssSetNumRWork(S, 7);
    ssSetNumIWork(S, 0);   /* integer work vec */
    ssSetNumPWork(S, 0);   /* pointer work vec */
    ssSetNumModes(S, 0);
    ssSetNumNonsampledZCs(S, 0);
    ssSetOptions(S, 0);
}
/* *********************************************************/
/*              mdlInitializeSampleTimes                  */
/* *********************************************************/
static void mdlInitializeSampleTimes(SimStruct *S)
{
   double T_samp ;
   /* Notice that we must get the sample time parameter  */
   /* directly - we can't use RWork yet.                 */
   T_samp = *mxGetPr(ssGetSFcnParam(S,5)) ;
   ssSetSampleTime(S, 0, T_samp);
   ssSetOffsetTime(S, 0, 0.0);
}
/* *********************************************************/
/*                   mdlInitializeConditions              */
/* *********************************************************/
#define MDL_INITIALIZE_CONDITIONS
static void mdlInitializeConditions(SimStruct *S)
{
   real_T *x0 ;
   x0 = ssGetDiscStates(S) ;
   *x0 = 0.0 ;
   ssSetRWorkValue(S, 0, 0.0) ; /* Initialize to zero */
}
#define MDL_START
/* *********************************************************/
/*                       mdlStart                         */
/* *********************************************************/
static void mdlStart(SimStruct *S)
  {
```

```
    mdlProcessParameters(S) ;
    }
/* ********************************************************* */
/*                mdlCheckParameters                         */
/* ********************************************************* */
#define     MDL_CHECK_PARAMETERS
static void mdlCheckParameters(SimStruct *S)
    {
    if (*mxGetPr(ssGetSFcnParam(S,0)) < 0)
        ssSetErrorStatus(S, "Proportional gain must be positive");
    if (*mxGetPr(ssGetSFcnParam(S,3)) < 0)
        ssSetErrorStatus(S, "Upper saturation limit must be positive");
    if (*mxGetPr(ssGetSFcnParam(S,4)) > 0)
        ssSetErrorStatus(S, "Lower saturation limit must be negative");
    }
/* ********************************************************* */
/*                     mdlProcessParameters                  */
/* ********************************************************* */
#define MDL_PROCESS_PARAMETERS
static void mdlProcessParameters(SimStruct *S)
    {
    int_T i;
    for(i = 0 ; i < 6 ; i++)
        {
        ssSetRWorkValue(S, (i + 1), *mxGetPr(ssGetSFcnParam(S,i))) ;
        }
    }
/* ********************************************************* */
/*                        mdlOutputs                         */
/* ********************************************************* */
static void mdlOutputs(SimStruct *S, int tid)
{
    double Kp, Ki, Kd, T_samp, u_prev, u_deriv ;
    real_T *y, *x ;
    int_T i ;
    InputRealPtrsType uPtrs ;
    double u ;
    uPtrs = ssGetInputPortRealSignalPtrs(S,0);
    u = *uPtrs[0] ;
    y = ssGetOutputPortRealSignal(S, 0) ;
    x = ssGetDiscStates(S) ;
    Kp = Kp_macro ;
    Ki = Ki_macro ;
    Kd = Kd_macro ;
    T_samp = T_samp_macro ;
    u_prev = ssGetRWorkValue(S, 0) ;
    ssSetRWorkValue(S, 0, u) ; /* Reset to current input */
    u_deriv = (u - u_prev)/T_samp ;
    *y = Kp*u + Ki*(*x) + Kd*u_deriv ;
}
/* ********************************************************* */
/*                        mdlUpdate                          */
/* ********************************************************* */
#define MDL_UPDATE
static void mdlUpdate(SimStruct *S, int_T tid)
{
    real_T *y, *x ;
```

```
    double T_samp, L_upper, L_lower ;
    InputRealPtrsType uPtrs ;
    double u ;
    uPtrs = ssGetInputPortRealSignalPtrs(S,0);
    u = *uPtrs[0] ;
    x = ssGetDiscStates(S) ;
    T_samp = T_samp_macro ;
    L_upper = L_upper_macro ;
    L_lower = L_lower_macro ;
    *x = *x + T_samp*u ;            /* Compute update */
    if(*x > L_upper) *x = L_upper ;  /* Apply limits */
    if(*x < L_lower) *x = L_lower ;
}
/* ******************************************************/
/*                       mdlTerminate                 */
/* ******************************************************/
static void mdlTerminate(SimStruct *S)
{
}
#ifdef MATLAB_MEX_FILE
#include "simulink.c"
#else
#include "cg_sfun.h"
#endif
```

10.6   SUMMARY

In this chapter, we described using S-Functions to build custom Simulink blocks. We discussed the functions each S-Function must perform and the general structure of an S-Function. Next, we discussed M-file S-Functions in detail, and presented examples using S-Functions with no states, with continuous states, and with discrete states. Last, we discussed C S-Functions and presented similar examples. Once you master the basics of S-Function programming, you will find additional useful information in the simstruc.h file, the online Simulink documentation, and Reference [3].

REFERENCES

[1] Dabney, James B. and Harman, Thomas L., *Mastering Simulink 2* (Upper Saddle River, N.J.: Prentice Hall, 1998).

[2] *Application Programming Interface Guides*, Version 6 (Natick, Mass.: The MathWorks, Inc., 2000).

[3] *Writing S-Functions*, Version 4 (Natick, Mass.: The MathWorks, Inc., 2000).

11

Graphical Animations

In this chapter, we will discuss four methods you can use to add graphical animations to a Simulink model. First, we'll discuss using MATLAB Handle Graphics in an S-Function to build an animation. Next, we will show how to use the Animation Toolbox. Finally, we will briefly discuss the Dials & Gauges Blockset and the Virtual Reality Toolbox.

11.1   INTRODUCTION

Graphical animations can make it easier to visualize a process being simulated. For example, one of the demonstrations included with Simulink is a model of a cart with an inverted pendulum. As the simulation progresses, the cart and pendulum move to graphically depict the state of the simulation. This animation makes it much easier to visualize the motion of the system.

In Chapter 9, we showed how you can build graphical animations using callbacks. This chapter presents four alternatives that are easier to use: S-Function-based animations, the Animation Toolbox, the Dials & Gauges Blockset, and the Virtual Reality Toolbox. Each of these methods of building graphical animations has advantages and disadvantages. The callback approach discussed earlier is the most powerful of these techniques, as it permits you to include custom controls in the animation window, and also allows you to build an animation block that can be open or closed during a simulation. An S-Function-based animation is easier to build but doesn't conveniently support adding custom controls. The Animation Toolbox is the easiest and fastest method of building a graphical animation. The Animation Toolbox does not allow you to add custom controls to the animation window, and it requires some additional effort on the part of the user. The Dials & Gauges Blockset makes it easy to add attractive input and output devices but does not provide any capability for more general animations. The Virtual Reality Toolbox provides a capability similar to callback-based animations but also allows you to develop the graphics using special graphical editors compatible with Virtual Reality Modeling Language.

11.2   S-FUNCTION ANIMATIONS

S-Function-based animations are S-Functions that have no states and no outputs. Therefore, they are custom sinks. An animation S-Function has two main parts: initialization and update. During initialization, the figure window is created, and the animation objects are prepared. During updates, the properties of the animation objects are changed to cause the objects to move or change in some other way as a function of the S-Function block input.

11.2.1   Animation S-Function Initialization

The initialization phase of an S-Function animation consists of S-Function initialization, as discussed in Chapter 10, and figure initialization. The sample time should be set to some small value such that the animation is acceptably smooth but that does not cause the simulation to run too slowly. You can find a suitable value with a little experimentation. A good initial guess might be (*Start time* − *Stop time*)/100. Don't set the sample time to 0 (continuous system), because it will cause the animation to appear erratic.

It is necessary to have some mechanism to determine whether an animation figure associated with the current S-Function block is already open. The approach used here is to store the path to the current block (from the gcb command) in the figure UserData parameter. It is safe to use gcb here because during execution of an S-Function, gcb will always return the path to the S-Function block. The figure initialization logic starts by determining whether the figure associated with the current block is already open. Suitable MATLAB statements are:

```
if(isempty(findobj('UserData',gcb)))

% initialize figure

end
```

If the figure is not already open, create it using the figure statement, for example,

```
h_fig = figure('Position',[x_pos, y_pos, width, height], ...
```

Next, store the path to the current S-Function block in the figure UserData so that the test for existence of the figure just described will work correctly. The following statement will set the UserData:

```
set(h_fig,'UserData',gcb) ;
```

Next, draw the animation figure using MATLAB graphics commands. (For a good tutorial on MATLAB graphics, see Hanselman and Littlefield [3].) Save the handles of graphics elements that will move or change in some other way as the simulation progresses. For example, suppose an element is defined by vectors x_array and y_array. Plot the points, and save the handle using the statement

```
hd1 = plot(x_array, y_array) ;
```

The last step in figure initialization is to save the handles of the elements that will change during the simulation. The technique used here to save these elements is to group them into a single MATLAB variable, and store that variable in the S-Function block UserData. The UserData can store any MATLAB variable, including cell arrays and structures, so there is no restriction on the type or amount of working data you can store. Suppose there are two graphics elements that are to be animated, and their handles are named hd1 and hd2. The following statements will store them in a structure:

```
t_data.hd1 = hd1 ;

t_data.hd2 = hd2 ;

set_param(gcb,'UserData',t_data) ;
```

11.2.2   Animation S-Function Updates

The animation S-Function will be treated as a discrete block, because we set the sample time to a positive number. Simulink will execute the S-Function at the sample times with flag = 2. The update function should first read from the S-Function block UserData the handles of the graphics objects that will change. For example, if the handles are stored in a structure variable as indicated above, they can be read as follows:

```
t_data = get_param(gcb,'UserData') ;

hd1 = t_data.hd1 ;

hd2 = t_data.hd2 ;
```

Next, compute new values for the properties of the objects that change, and update the properties using the set command:

```
set(handle, PropertyName, PropertyValue) ;
```

where *handle* is the handle of the object, PropertyName is a MATLAB string containing the name of the object property to be changed, and PropertyValue is the new property value.

■
_____

EXAMPLE 11.1   Rolling disk animation

In this example, we will build a graphical animation of a disk rolling back and forth in a semicircular trough. Figure 11.1 illustrates the problem.

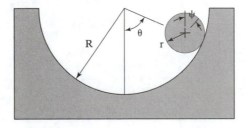

FIGURE 11.1: Disk rolling in a trough

Assuming the disk rolls without slipping, the equation of motion of this system is

$$\ddot{\theta} = \frac{-g\sin\theta}{1.5\,(R - r)}$$

where g is the acceleration due to gravity. The kinematic relationship between ψ and θ is

$$\psi = \theta\,\frac{R - r}{r}$$

Figure 11.2 illustrates a Simulink model of the equation of motion of this system, with an animation S-Function block to show the motion of the disk. We assume that $R = 12, r = 2$, and $g = 32.2$.

The animation S-Function is shown in Listing 11.1, and the animation figure is shown in Figure 11.3.

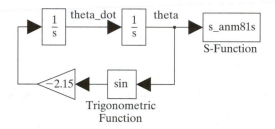

FIGURE 11.2: Disk rolling in a trough model

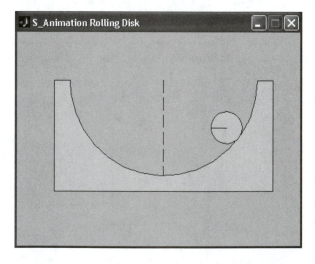

FIGURE 11.3: Disk rolling in a trough animation

Listing 11.1: Rolling disk animation S-Function

```matlab
function [sys,x0,str,ts] = s_anm81s(t,x,u,flag)
%S-Function animation example 1
%This example demonstrates building an animation
%using a single S-Function with no callbacks.
%
%Based on sfuntmpl.m, supplied with SIMULINK
%Copyright (c) 1990-2002 by The MathWorks, Inc.
%
switch flag,
  case 0,                  %Initialization
    [sys,x0,str,ts]=mdlInitializeSizes;
  case 2,
    sys=mdlUpdate(t,x,u);
  case 3,                  %Outputs
    sys = mdlOutputs(t,x,u) ;
  case 9,                  %Finished. Do any needed
    sys=mdlTerminate(t,x,u);
  otherwise               %Invalid input
    error(['Unhandled flag = ',num2str(flag)]);
end


%*********************************************************
%*                  mdlInitializeSizes                   *
%*********************************************************
function [sys,x0,str,ts]=mdlInitializeSizes()
% Return the sizes of the system vectors, initial
% conditions, and the sample times and offsets.
sizes = simsizes;   % Create the sizes structure
sizes.NumContStates  = 0;
sizes.NumDiscStates  = 0;
sizes.NumOutputs     = 0;
sizes.NumInputs      = 1;
sizes.DirFeedthrough = 0;
sizes.NumSampleTimes = 1;
sys = simsizes(sizes);
x0  = [];        %There are no states.
str = [];        %str is always an empty matrix.
                 %Update the figure every 0.25 sec.
ts  = [0.25 0]; %Initialize the array of sample times.


% Initialize the figure.
% The handles of the disk and index mark are stored
% in the block's UserData.
if(isempty(findobj('UserData',gcb)))
  h_fig = figure('Position',[200 200 400 300], ...
                 'MenuBar','none','NumberTitle','off', ...
                 'Resize','off', ...
                 'Name',[gcs,' Rolling Disk']) ;
  set(h_fig,'UserData',gcb) ; %Save name of current block
                              %in the figure's UserData.
                              %This is used to detect
                              %that a rolling disk figure
                              %is already open for the
                              %current block, so that
```

```
                                    %only one instance of the
                                    %figure is open at a time
                                    %for a given instance of the
                                    %block.
  set(h_fig,'DoubleBuffer','on') ;
  r = 2 ;
  R = 12 ;
  q = r ;
  thp = 0:0.2:pi ;
  xp = R*cos(thp);
  yp = -R*sin(thp) ;
  xp = [xp,-R,-(R+q),-(R+q),(R+q), (R+q),R] ;
  yp = [yp,0,0,-(R+q),-(R+q),0,0] ;
  cl_x = [0,0] ;
  cl_y = [0,-R] ;
  % Make the disk
  thp = 0:0.3:2.3*pi ;
  xd = r*cos(thp);
  yd = r*sin(thp) ;
  hd = fill(xp,yp,[0.85,0.85,0.85]); %Draw trough
  hold on ;                          %so it won't get erased.
  set(hd,'EraseMode','none');
  axis('equal');axis('off');
  hd0 = plot(cl_x,cl_y,'k--');    %Draw the centerline.
  set(hd0,'EraseMode','none');
  %During this initialization pass, create the disk
  %(hd2) and the index mark (hd3).
  theta = 0 ;
  xc = (R-r)*sin(theta);    %Find center of disk.
  yc = - (R-r)*cos(theta) ;
  psi = theta*(R-r)/r ;
  xm_c = r*sin(psi) ;    %Relative position of index mark.
  ym_c = r*cos(psi) ;
  xm = xc + xm_c ;       %Translate mark.
  ym = yc + ym_c ;
  hd2 = fill(xd+xc, yd+yc, ...
             [0.85,0.85,0.85]) ; % Draw disk and mark.
  hd3 = plot([xc,xm],[yc,ym],'k-');
  set_param(gcb,'UserData',[hd2,hd3]) ;
end


%***********************************************************
%*                    mdlUpdate                           *
%***********************************************************
function sys=mdlUpdate(t,x,u)
% Compute update for discrete states.
sys = [];    % Empty because this model has no states.
% Update the figure.
r = 2 ;
R = 12 ;
q = r ;
userdat = get_param(gcb,'UserData') ;
hd2 = userdat(1) ;
hd3 = userdat(2) ;
theta = u(1) ;       %The sole input is theta.
xc = (R-r)*sin(theta);    %Find center of disk.
yc = - (R-r)*cos(theta) ;
```

```
psi = theta*(R-r)/r ;
xm_c = r*sin(psi) ;   %Find relative position of index.
ym_c = r*cos(psi) ;
xm = xc + xm_c ;      %Translate mark.
ym = yc + ym_c ;
thp = 0:0.3:2*pi ;
xd = r*cos(thp);
yd = r*sin(thp) ;
%Move the disk and index marks to new positions.
set(hd2,'XData',xd+xc);
set(hd2,'YData',yd+yc);
set(hd3,'XData',[xc,xm]);
set(hd3,'YData',[yc,ym]);


%************************************************************
%*                      mdlOutput                         *
%************************************************************
function sys=mdlOutputs(t,x,u)
%Compute output vector.
sys = [];


%************************************************************
%*                    mdlTerminate                        *
%************************************************************
function sys=mdlTerminate(t,x,u)
%Perform any necessary tasks at the end of the simulation.
sys = [];
```

11.3   ANIMATION TOOLBOX

The Animation Toolbox is an extension to Simulink that permits you to create an animation graphically, in a manner similar to building a simulation in Simulink. To build an animation, you copy graphic objects to a figure window, and then set properties of the objects using object dialog boxes.

11.3.1   Obtaining the Animation Toolbox

The Animation Toolbox is available on The MathWorks Web site in the `File Exchange` area under the `Simulation and Modeling` category. The toolbox is free, but it requires Simulink. To install the Animation Toolbox, first download it to your computer using a Web browser or file transfer protocol (FTP) program. Add the directory in which you stored the toolbox files to the top of the MATLAB path so that Simulink can find the toolbox files.

11.3.2   Using the Animation Toolbox

To add an Animation block to a model, open the Simulink model `animblk.mdl` in the Animation Toolbox directory, and drag the Animation block to your model window. The Animation block can accept scalar or vector inputs. Double-click the Animation block, opening the animation figure window shown in Figure 11.4.

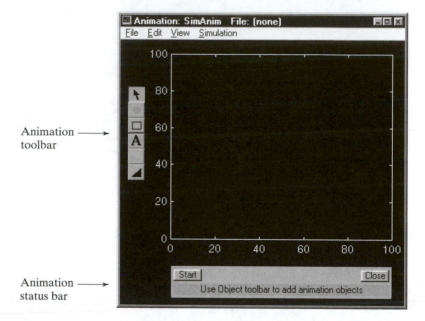

FIGURE 11.4: Animation Toolbox figure window

The animation figure window consists of three areas: the Animation Toolbar, the Animation Status bar, and the animation figure. The Animation Toolbar consists of a set of icons representing the available animation objects: dot, rectangle, text, line, and patch. The Animation Status bar displays the animation status and provides buttons to control the simulation and close the animation figure. When the simulation is running, the Animation Status bar contains a **Stop** button, a check box to show trails, and a button to clear the trails. The animation figure is the area in which the animation is created and displayed.

To construct an animation, click one of the icons in the Animation Toolbar, then click the animation figure at the desired location of the object represented by the icon. An object dialog box will pop up. Enter the object configuration data in the dialog box, then choose **Apply**, and close the dialog box. Repeat for each object in the animation. Use the various choices on the **View** menu to configure the display. You can run the simulation using the Animation Status bar **Start** button or **Simulation:Start** from the model window menu bar.

■
EXAMPLE 11.2   Using the animation toolbox

To illustrate the process of building an animation using the Animation Toolbox, consider the Simulink model shown in Figure 11.5. The model consists of two Sine Wave blocks connected to a Mux to produce a vector signal. The lower Sine Wave block is configured with **Phase** set to pi/2.

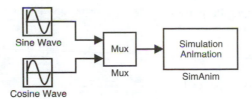

FIGURE 11.5: Simulink model with Animation block

Double-click the Animation block, opening the animation figure window.

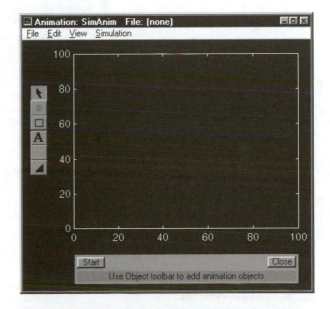

Click the dot icon, then click the figure.

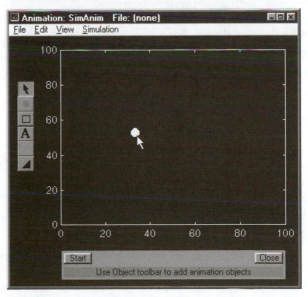

The Dot Object **Properties** dialog box will automatically be displayed. Configure the block dialog box fields as shown, then press **OK** and close the dialog box.

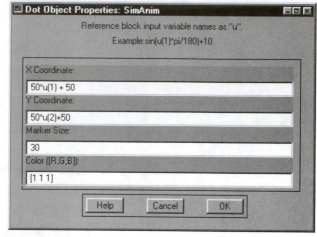

Run the simulation by clicking on the **Start** button. The dot should move in a circle.

11.3.3   Animation Object Properties

Each of the properties fields in the object dialog boxes can contain expressions consisting of constants, variables currently defined in the MATLAB workspace, and elements of the input vector (u). In addition to making the location of an object a function of the input, you can also make the color or size of an object a function of the input.

11.3.4   Configuring an Animation

The **View** menu contains several options that are useful in tailoring the appearance of an animation figure.

Figure Properties.   The **View** menu provides options to show or hide several figure properties. You can turn the axes on or off, turn a grid on or off, and display or hide a border. There are also options to show or hide the Status bar and the Toolbar.

Figure Scale.   **View:AutoScale** allows you to control whether or not the animation figure automatically scales. The autoscale feature can be useful in determining appropriate axis limits. Once you've found acceptable limits, the appearance of the animation can be improved by setting **AutoScale** to off and then using the **View:Change Axis Limits** to fine-tune the limits.

11.3.5   Modifying an Animation

After running a simulation, you can add objects to an animation figure or modify the objects in the figure. Before modifying the figure, choose **Simulation:Reset** from the animation window menu bar. To modify an existing object, double-click the object.

11.3.6   Setting Initial Inputs

You can set the initial values of animation block input variables [u(1),u(2), etc.] by selecting **Simulation:Set Initial Inputs**. This will open the dialog box shown in Figure 11.6. Enter the initial values of the inputs as a vector. This capability is useful, because the Animation block does not have access to the inputs until the simulation begins. Initializing the inputs prevents a large start-up transient in the animation figure. Setting the inputs in this dialog box has no effect on the inputs; these values do not propagate backwards from the Animation block. The initial values default to zero.

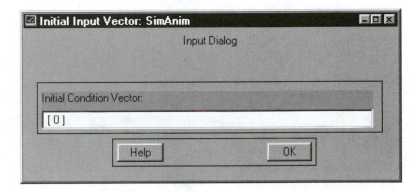

FIGURE 11.6: Animation initial inputs dialog box

11.3.7   Saving and Loading Animation Files

The File menu of the animation figure window provides options to save or load an animation. The information that the Animation Toolbox needs to generate an animation figure is stored in the form of a MATLAB .mat file. When you open an animation figure window by double-clicking an Animation block, and if the last time you accessed that block you saved an animation associated with that block, there will be a prompt asking whether you wish to use the previous animation or start a new one.

11.4   DIALS & GAUGES BLOCKSET

Among the many optional Simulink toolboxes is the Dials & Gauges Blockset. This toolbox is a collection of source (dials) and sink (gauges) blocks that represent a variety of commonly used input and display devices. The Dials & Gauges Blockset is available only for the Windows environment. It employs the ActiveX mechanism for data communication between Simulink and the blocks. The toolbox is particularly useful for developing control or instrument panel layouts and performing usability studies.

When the Dials & Gauges Blockset is installed, it appears as a block library in the Simulink Library Browser. The dials and gauges may be displayed as part of the Simulink block diagram by installing them in the Simulink model in a manner similar to conventional blocks. The library also includes a communication mechanism that allows you to place the dials and gauges in a subsystem, in a separate Simulink model, or even in a MATLAB figure.

We will illustrate the use of the blockset by modifying the automobile speed control model of Chapter 7 with a speed control knob and a speedometer gauge. Details for the other implementation options may be found in the user's guide [2].

Starting with a copy of the automobile speed control model, remove the constant source and slider gain, and replace with a Generic Knob block. Also add a Generic Angular Gauge block.

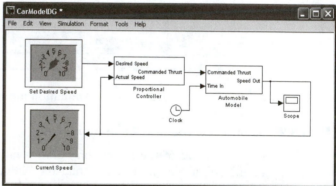

Flip the Generic Angular Gauge block, and resize the Generic Knob block. Connect and label the blocks as shown.

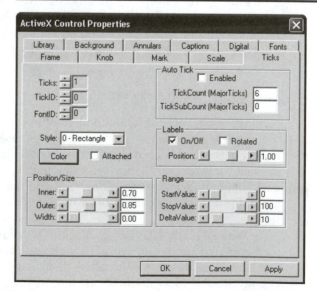

Double-click the border of the Generic Knob block to open the block dialog box. Note that you must double-click the border rather than the middle of the block. Select the **Scale** page and set field **Min** to 0 and **Max** to 100.

Select the **Ticks** page, and set **StartValue** to 0, **StopValue** to 100, and **DeltaValue** to 10. Click OK.

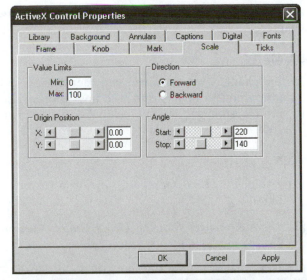

Configure the Generic Angular Gauge block similarly. The model now appears as shown.

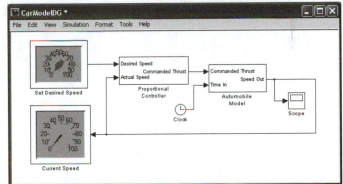

The blocks in the Dials & Gauges Blockset have many configuration options. You can set colors, change scales, add captions, and customize the appearance of all the graphical elements.

11.5 VIRTUAL REALITY TOOLBOX

The Virtual Reality Toolbox provides a mechanism to connect Simulink models to graphical animations written in Virtual Reality Modeling Language (VRML). VRML provides object-oriented graphics capabilities similar to MATLAB Handle Graphics. The animations are displayed using a VRML-enabled Web browser. An advantage of VRML is that there are a number of graphics editors that facilitate the development of the objects and backgrounds. The Virtual Reality Toolbox is furnished with graphics editor V-Realm Builder from Ligos Corporation, San Francisco, California.

The interface between Simulink and the VRML animation is implemented in blocks included with the toolbox. An animation is then developed in VRML code directly or indirectly using the graphics editor. The VRML animation receives signals from Simulink and changes properties of graphical elements based on the signals in a manner similar to the MATLAB animations we discussed earlier in this chapter.

11.6 SUMMARY

In this chapter, we explained two methods for adding graphical animations to Simulink models. First, we showed how you can use S-Function-based animations to create custom animation blocks. We then explained how to use the Animation Toolbox, which is available from The MathWorks Web site.

REFERENCES

[1] *Using MATLAB Graphics*, Version 6 (Natick, Mass.: The MathWorks, Inc., 2000).

[2] *Dials & Gauges Blockset User's Guide*, Version 1 (Natick, Mass.: The Math-Works, Inc., 2002).

[3] Hanselman, Duane C., and Littlefield, Bruce R., *Mastering MATLAB 6: A Comprehensive Tutorial* (Upper Saddle River, N.J.: Prentice Hall, 2001).

12

Debugging

In this chapter, we will discuss debugging Simulink models. We will discuss finding and eliminating model-building (syntax) errors and implementation errors. We'll also describe the Simulation Diagnostics dialog box and two debugging tools: floating output (Scope and Display) blocks and the Simulink debugger.

12.1 INTRODUCTION

Programming in Simulink is generally less error-prone than programming in traditional text-based programming languages, because the higher level of abstraction reveals the content and structure of the program more clearly. But of course, Simulink programming is still not error-free, so it's important to have tools and techniques for identifying and correcting programming errors. In this chapter, we will consider two classes of programming errors: model-building (syntax) errors and implementation errors.

There is a third important class of errors that can be called algorithmic. This type of error arises when the model is built in accordance with its design, but the design does not achieve its objectives. Algorithmic errors are not normally uncovered in debugging. Instead, they are usually discovered during the process of model validation testing.

12.2 FINDING MODEL-BUILDING ERRORS

A model-building error is a characteristic of a model that prevents it from executing as intended, corresponding roughly to a syntax error in a traditional programming language. Examples of model-building errors are unconnected signal lines, unconnected block inputs and outputs, and missing block configuration data. Simulink can automatically detect each of these errors during the compilation step when the model is run, and report them as errors via the Simulation Diagnostics dialog box or via warning messages displayed in the MATLAB window.

To ensure that you locate all model-building errors during model development, you can configure the appropriate **Simulation:Parameters Diagnostics** page options Unconnected line, Unconnected block input, and Unconnected block output to Error, as recommended in Section 4.10.3. When you run the simulation, Simulink will check for each type of error, and, if any are found, the simulation will stop, and the Simulation Diagnostics dialog box will appear. If Simulink detects a model-building error for which the diagnostic option is **Warning**, a warning message will be displayed in the MATLAB window, but the simulation will proceed.

To illustrate the Simulation Diagnostics dialog, consider the Simulink model shown in Figure 12.1. The **Simulation:Parameters Diagnostics** page options for model building errors are set to **Error**. When the model is run, the Simulation Diagnostics dialog box appears as shown in Figure 12.2. The errors are listed in compilation order, and the first block in the list is highlighted as shown in Figure 12.3.

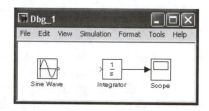

FIGURE 12.1: Model with missing signal line

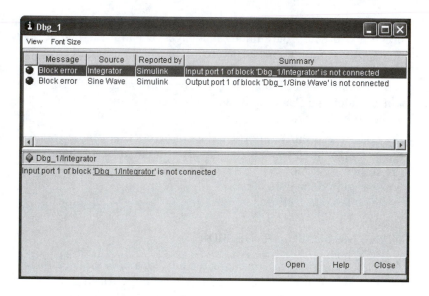

FIGURE 12.2: Simulation Diagnostics dialog box

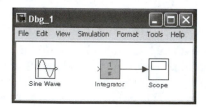

FIGURE 12.3: Simulink model with block highlighted

The Simulation Diagnostics dialog box (Figure 12.2) consists of three sections. The top section is a table of four columns and a row for each diagnostic message. The first column defines the type of message for the row. The second column lists the type of block associated with the message. The third column lists the source of the message, normally Simulink. The fourth column lists an abbreviated diagnostic message.

The third column lists the full Simulink path to the block associated with the message.

The second section of the Simulation Diagnostics dialog box is the message window. The first line in this window lists the full Simulink path to the block associated with the error. The following lines display a detailed diagnostic message corresponding to the abbreviated message in the fifth column of the highlighted row in the top section. Initially, the top row of the table in the top section is highlighted. To highlight another row, click anywhere in the row.

The bottom section of the Simulation Diagnostics page contains three buttons, **Open**, **Help**, and **Close**. Press **Open** to highlight the block associated with the highlighted row. Press **Close** to dismiss the Simulation Diagnostics dialog box.

12.3   FINDING IMPLEMENTATION PROBLEMS

An implementation problem is an undesirable model characteristic that may not prevent the model from executing but prevents it from correctly and efficiently modeling the dynamical system. Some implementation problems will cause Simulink execution errors, such as floating-point overflow (if the **Simulation:Parameters** diagnostic is set to **Error**). Other implementation problems, such as chatter due to zero crossings and algebraic loops, can cause the simulation to execute very slowly. The most difficult type of implementation problem to identify is one that does not cause any apparent simulation anomaly but causes the simulation to produce incorrect results. Examples of this type of error are sign errors and connecting the wrong signal line to a block input. We will discuss each of these types of errors next.

12.3.1   Simulink Execution Errors

Simulink traps and reports many execution errors via the Simulation Diagnostics dialog box discussed in the previous section. Unless there is some reason that data overflow is expected behavior, the **Simulation:Parameters** diagnostics should be configured to treat data overflow as an error. Using the Simulation Diagnostics dialog box, it is usually fairly easy to identify and correct execution errors.

12.3.2   Slow Execution

A Simulink model can execute very slowly for a variety of reasons. Among these are zero-crossing chatter, algebraic loops, stiffness, and model complexity. The best approach to debugging is to check for one type of problem at a time.

Zero-Crossing Chatter.   To determine whether zero-crossing chatter is causing the simulation to run slowly, set **Simulation:Parameters Advanced** page optimization Zero crossing detection to **Off**. If the simulation speeds up noticeably, zero-crossing chatter is occurring.

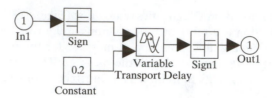

FIGURE 12.4: Delayed Sign subsystem

The most common cause of zero-crossing chatter is a block that exhibits a discontinuity at a zero crossing. When a variable-stepsize solver is used for such a model, the solver will adjust the stepsize, perhaps to a very small value, so as to locate the zero crossing. If the zero crossings occur infrequently, this is not much of a problem. On the other hand, if the signal remains near zero, the solver will locate each zero crossing as closely as possible, consuming a lot of time. This chatter is usually an artifact of the model, not the dynamical system. For example, in a switching controller modeled using a Sign block, there must be a finite switching time. Replacing the Sign block with the delayed Sign subsystem shown in Figure 12.4 will more accurately model the physical system and significantly reduce the modeling chatter.

Algebraic Loops.   An algebraic loop is a model characteristic in which the input of a block is an algebraic function of the output of the same block. Algebraic loops are discussed in detail in Section 13.3. To detect algebraic loops, set **Simulation:Parameters** Diagnostics item `Algebraic loop` to **Error**. If an algebraic loop is present, the Simulation Diagnostics dialog box will identify and report the problem.

Stiffness.   Stiffness is usually not an implementation error, although it can cause a simulation to execute extremely slowly. Section 13.2 discusses stiff systems in detail. The best way to determine if a system is executing slowly due to stiffness is to try the various stiff solvers. If the stiff solvers cause the model to execute at an acceptable speed, the problem is probably solved.

Model Complexity.   Generally speaking, the more complex a model is, the slower it will execute. While it may not be practical to eliminate the complexity in a model, it is useful to know which parts of a model consume the most time, and therefore, which will benefit most from efficiency improvements. The MATLAB profiler produces a report that provides this information. To use the profiler, enter the command

```
profile viewer
```

at the MATLAB prompt. The MATLAB profiler window (Figure 12.5) will be displayed. In field **Run this code**, enter a suitable `sim` command (see Chapter 8). The profiler will then execute the model and produce a report that is displayed in the profiler window.

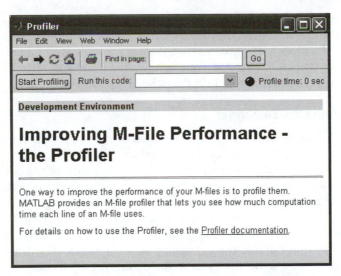

FIGURE 12.5: MATLAB profiler

12.3.3   Incorrect Results

An important step in testing software is to determine the expected results of each test case before the case is run, so that the simulation results can be evaluated quantitatively. If the simulation results differ from the expected results, debugging is necessary. Simulink provides a set of useful tools to assist in debugging. Two that we will discuss here are the floating Scope block and the floating Display block. In the next section, we will discuss the Simulink debugger.

Floating Scope Block.   A floating Scope block is a Scope block that has no input signal. A floating Scope block can be used as an oscilloscope to inspect any signal line in the model during a simulation. To use a floating Scope block, you should first disable **Simulation:Parameters Advanced** page option `Signal storage reuse`. Next, copy a floating Scope block to the model window. Now you can inspect any signal line in the model by selecting the signal line before or during a simulation. The floating Scope block will display the selected signal just as if the signal were connected to the Scope block input port.

Floating Display Block.   A floating Display block has no input signal. The block can be used in the same manner as the floating Scope block. To use a floating Scope block, you should first disable **Simulation:Parameters Advanced** page option `Signal storage reuse`. Copy a Display block to the model window. Open the block dialog box and select check box **Floating display**.

12.4   SIMULINK DEBUGGER

An interactive debugger is a software tool that allows you to control the execution of a computer program, pausing execution at selected times or when certain events occur. An interactive debugger also allows you to monitor the values of program variables. The Simulink debugger provides many of the features of a conventional interactive debugger. It allows you to execute a model block by block, one time step at a time, or continuously until some event occurs. The Simulink debugger also allows you to inspect the value of any signal in a model. In this section, we will describe running a model via the debugger and several useful debugger commands.

12.4.1 Running the Debugger

The Simulink debugger can be operated from the MATLAB command line prompt or the debugger graphical user interface (GUI). The debugger GUI interface is easier to use and provides most of the functionality of the command line interface, so we will concentrate on the GUI interface here. To debug a model, open the model, and then start the debugger by choosing **Tools:Debugger** from the model window menu bar or click the debug icon on the model window toolbar. The debugger GUI will be displayed as shown in Figure 12.6. The debugger toolbar contains seven buttons that control the debugging process and a help button. The toolbar also contains a Close button to close the debugging GUI. The buttons are described in Table 12.1. When the debugger starts, the model window will expand to show the model browser pane.

The debugger window contains three panes. The Output pane shows debugger output and also has tabs to show the block execution order and debugger status. The Break/Display points pane lists blocks selected as breakpoints or display points.

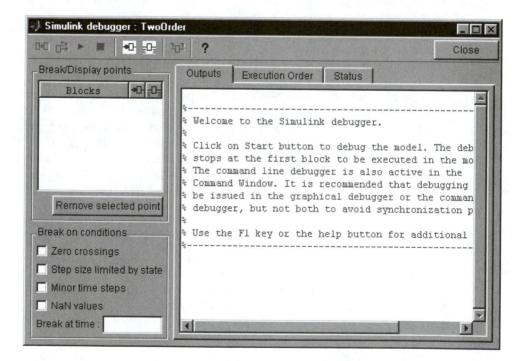

FIGURE 12.6: Debugger GUI window

TABLE 12.1: Debugger Command Buttons

Button	Purpose
	Step to next block Causes the debugger to step to the next block in the evaluation sequence.
	Go to start of next time Causes the debugger to proceed to the next time step.
	Start Initiates the debugging process and stops at the first block. Also causes the debugger to proceed when it's halted.
	Stop Stop the current simulation. The debugger remains in operation.
	Breakpoint Set a breakpoint before the current block.
	Display point Set a display point at the current block.
	Show block I/O Display the input and output signals of the selected block each time the simulation pauses.
	Help Open the help browser to the debugger page.

The `Break on conditions` pane contains a set of check boxes to control condition breakpoints and a field to enter a time breakpoint. We will discuss each part of the debugger window as we explain the use of the debugger.

To start debugging, click the **Start** button. The first block in the execution order will be highlighted, and the debugger output pane will display a starting message:

```
[Tm = 0 ] **Start** of system 'TwoOrder' outputs

%-------------------------------------------------------%

(sldebug @0:0 'TwoOrder/Constant'):
```

Once the debugger starts, you interact with it using the debugger toolbar buttons, entering commands at the MATLAB command line, and clicking blocks in the model window.

12.4.2   Block Index

Although a Simulink model is in the form of a block diagram, which implies continuous parallel evaluation of all blocks, the mechanization of a model requires that the blocks be evaluated sequentially, with certain blocks involved in algebraic loops evaluated iteratively each time step. Simulink maintains a list of all blocks in a model, sorted in execution order. The execution order is a two-number code in the form s:b. Here, s is the subsystem number, and b is the block number within the subsystem, with numbering starting at zero. Thus, the third block in the second subsystem would be block 1:2. To see a complete block inventory, click the **Execution Order** tab in the debugger output pane.

A number of the Simulink debugger commands available from the MATLAB prompt require as an argument a block index. Simulink recognizes two conventions to identify a block for these commands: block number and command gcb (with the desired block currently selected). For example, to display the inputs and outputs to block 0:1 which is currently selected, the two following commands are equivalent:

```
probe 0:1
```

```
probe gcb
```

In the discussion of the various debugger commands, we will use the term *index* to indicate the argument to any command that requires a block index. So the probe command will be shown as probe *index*.

12.4.3 Single-Step Execution

The Simulink debugger provides commands that allow you to single step through model execution. You can step either block by block or by time step. Block-by-block stepping is performed in the sorted block order. To step block by block, click the **Step by block** button once for each step.

If you step block by block, the debugger displays a block report for the current block and highlights the block in the model window. A block report lists the block index, the block number, and the values of the input and output signals. If continuous states are associated with the block (such as an Integrator block), the block report includes the value of the continuous state associated with the block. The input signals are identified as U1, U2, etc., and the output signals are identified as Y1, Y2, etc., and the continuous state is identified as CSTATE.

EXAMPLE 12.1 Single-step debugging

Consider the Simulink model shown in Figure 12.7 which we saved as TwoOrder.mdl. To step through the first time step block by block, start the debugger by choosing **Tools:Debugger** from the model window menu bar. Then click the debugger **Start** button. Click the **Step to next block** button. The debugger will proceed to the next block (here, 0:1, the first integrator), highlight the block, and display the block input (U1) and the block output (Y1) and continuous state (CSTATE). Press the **Step to next block** button repeatedly to step through the remaining blocks. After the last block executes, the time step report is displayed (the time is identified as TM). The debugging session is shown in Listing 12.1.

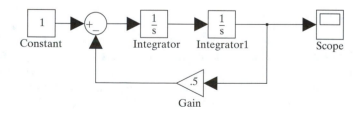

FIGURE 12.7: Model for debugging example

Listing 12.1: Step-by-block debugging

```
[Tm = 0                          ] **Start** of system 'TwoOrder' outputs
%-----------------------------------------------------------------%
(sldebug @0:0 'TwoOrder/Integrator1'):
U1 = [0]
CSTATE = [0]
Y1 = [0]
%-----------------------------------------------------------------%
(sldebug @0:1 'TwoOrder/Scope'):
U1 = [0]
%-----------------------------------------------------------------%
(sldebug @0:2 'TwoOrder/Constant'):
Y1 = [1]
%-----------------------------------------------------------------%
(sldebug @0:3 'TwoOrder/Gain'):
U1 = [0]
Y1 = [0]
%-----------------------------------------------------------------%
(sldebug @0:4 'TwoOrder/Integrator'):
U1 = [0]
CSTATE = [0]
Y1 = [0]
%-----------------------------------------------------------------%
(sldebug @0:5 'TwoOrder/Sum'):
U1 = [1]
U2 = [0]
Y1 = [1]
[Tm = 0.0002009509145207664  ] **Start** of system 'TwoOrder' outputs
%-----------------------------------------------------------------%
(sldebug @0:0 'TwoOrder/Integrator1'):
```

To step through a model by time steps, click the **Go to start of next time** button repeatedly.

12.4.4 Breakpoints

Often, it is desirable to pause the simulation upon reaching a condition of interest. The mechanism to cause the simulation to pause in this fashion is called a breakpoint. The Simulink debugger provides several types of breakpoints. A block breakpoint causes an execution to pause upon reaching a specified block. A time breakpoint causes the simulation to pause upon reaching a specified value of simulation time. An event breakpoint causes execution to pause upon occurrence of a specified event. To cause the simulation to proceed to the next breakpoint (or the end of the simulation if none are encountered), click the debugger **Start** button. To cause the simulation to proceed to the end, ignoring all breakpoints, enter the command run at the MATLAB prompt.

Setting Block Breakpoints.   To set a block breakpoint, select the block, and then click the **Breakpoint** icon. The breakpoint will be set such that the simulation stops just before the selected block is evaluated. Setting a breakpoint adds the

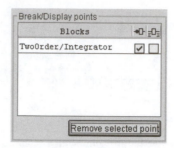

FIGURE 12.8: Break/Display points pane

block to the `Break/Display points` list pane with the **Breakpoint** check box selected. For example, setting a breakpoint at the first integrator block in the model shown in Figure 12.7 results in the `Break/Display points` list pane shown in Figure 12.8. You can disable the breakpoint by clicking the **Breakpoint** check box, or remove the breakpoint by selecting it in the list and then clicking **Remove display point**.

Setting Time Breakpoints. A time breakpoint causes the simulation to pause upon reaching a particular value of simulation time. You can set a single time breakpoint by entering the desired time in the field Break at time in the Break on conditions pane of the debugger window. To set multiple time breakpoints, enter the command `tbreak time` at the MATLAB prompt. If a time breakpoint is already set for that time, it is cleared. The command `tbreak` with no argument sets (or clears) a breakpoint at the current simulation time. So a useful sequence is to enter the command `tbreak time` to set the breakpoint, then click the debugger **Start** button to proceed to that breakpoint. After executing various debugger commands to investigate the state of the simulation at that time, enter `tbreak` again to clear the current time breakpoint.

Setting Event Breakpoints. Event breakpoints allow you to pause the simulation upon the occurrence of events that may cause erroneous results or slow execution. The event breakpoints are selected using the check boxes in the Break on conditions pane of the debugger window.

Select **Zero crossings** to cause the simulation to pause when a nonsampled zero crossing is detected. As discussed in Section 12.3, zero crossings can cause very slow simulation execution. Setting the zero-crossing breakpoint allows you to identify the block associated with the zero crossing. When the debugger detects a zero crossing with zero-crossing breakpoints enabled, the simulation pauses, and the affected block is highlighted.

Select **Step size limited by state** to help identify states that are causing the simulation to execute slowly. This is likely to occur in a stiff system. See Section 13.2 for a discussion of stiff systems.

Select **NaN values** to cause the simulation to pause when any signal value attains an undefined value, such as 0.0/0.0.

12.4.5 Inspecting Block Data

Three debugger commands entered at the MATLAB prompt control display of block reports. We already discussed the `trace` command, which causes the debugger to display a block report each time a block is executed. A similar command is `disp` *index*, which instructs the debugger to display a block report each time it pauses for a breakpoint.

The `probe` command allows you to display a block report for any block when the simulation is paused. The `probe` command has two forms. Enter `probe` *index* to display a block report for the specified block. Enter the command `probe` with no argument to activate probe mode. While probe mode is in effect, the debugger will display a block report in the MATLAB command window for any block you click. To cancel probe mode, enter any command.

12.4.6 Algebraic Loop Commands

The debugger provides two commands that can be entered at the MATLAB prompt to allow you to gain insight into solution of algebraic loops. Command `ashow` highlights algebraic loops in the Simulink model window. Command `atrace` displays algebraic loop solution information in the debugger window.

Simulink maintains a sorted list of algebraic loops using the notation *s#a*, where *s* represents the system, and *a* represents the number of the loop. Note that while the system number *s* starts numbering at zero, algebraic loop numbering for a system starts at one. Therefore, the first algebraic loop in the first system will be *0#1*.

To identify the blocks that compose an algebraic loop, use the command `ashow` *index*, where *index* is a block index (*s:b* or `gcb`) or algebraic loop index (*s#a*). The first form will highlight the algebraic loop containing the referenced block, assuming the block is in an algebraic loop. The second form will highlight the referenced algebraic loop. Use command `ashow clear` to remove algebraic loop highlighting.

Command `atrace` *level* controls algebraic loop tracing, where *level* is the trace level. Level 0 turns tracing off, level 1 displays the minimum amount of tracing information, and level 4 displays the maximum amount of tracing information. Table 12.2 lists the available trace levels and corresponding trace output.

TABLE 12.2: Trace Levels

Level	Tracing Output
0	Disables algebraic loop tracing.
1	Displays block index and path of the first block in the loop, the final value of the loop variable, the number of iterations, and an error estimate.
2	Same as Level 1.
3	Level 1 output plus the Jacobian matrix used to solve the algebraic loop.
4	Level 3 output plus convergence history for the algebraic loop variables.

12.4.7 Debugger Status Commands

Four debugger commands display information about the simulation and the current state of the debugger. Two of these commands are available from the debugger window, and the other two must be entered at the MATLAB prompt.

Block Inventory. A block inventory is produced by selecting the **Execution Order** tab of the debugger output pane. Certain blocks are not executable and therefore are not listed. For example, a Mux block is a signal-grouping device that makes a block diagram easier to read but does not change any signal's value. Therefore, a Mux block is not executable and is not part of the list produced by `slist`.

Debugger Status. The status of the debugger can be displayed at any time by clicking the **Status** tab of the debugger output pane. The status output consists of a list of current debugger settings. This list includes the current simulation time, state of options such as event breakpoints, and the algebraic loop tracing level.

Subsystem Inventory. Command systems, entered at the MATLAB prompt, display a list of Simulink subsystems in the MATLAB command window. This list only contains those subsystems that Simulink must treat as separate systems, termed nonvirtual systems. Ordinarily, subsystems are merely a means of programming abstraction that makes it easier to understand a model. As part of the compilation process, Simulink flattens the list of subsystems by expanding the subsystems as groups of blocks in the top level of the model. Certain subsystems, namely triggered and enabled subsystems, cannot be expanded into the top level of the model, and therefore, remain as nonvirtual systems.

Zero-Crossing Inventory. Command `zclist`, also entered at the MATLAB prompt, produces a list of blocks that may be associated with zero crossings during model execution. The fact that a particular block appears on this list does not guarantee that block will be associated with a zero crossing, it only guarantees that Simulink will monitor the block for zero crossing during a simulation.

12.4.8 Ending Debugging

The debugger ends and control returns to the MATLAB prompt when the simulation reaches its stop time. You can terminate debugging at any time by clicking the **Stop** button on the debugger toolbar.

12.4.9 Debugger Command Summary

Provided in Table 12.3 is a list of Simulink debugger commands that may be entered at the MATLAB prompt. The debugger recognizes a short form for each command that is equivalent to the full command, as indicated in the second column of the table. In the usage column, *index* indicates the block index, either in *s:b* form or via command gcb. The algebraic loop index must be in *s#a* form.

TABLE 12.3: Debugger Command Summary

Command	Short Form	Usage
ashow	as	ashow *textitindex* where *index* is either the block index or algebraic loop index. Highlights the referenced algebraic loop. ashow clear removes highlighting.
atrace	at	atrace *level* sets the algebraic loop trace level. *level* is an integer from 0 (no tracing) to 4 (detailed trace information).
bafter	ba	bafter *index* sets a breakpoint after the indicated block.
break	b	break *index* sets a breakpoint before the indicated block.
bshow	bs	bshow *s:b* selects the indicated block. Equivalent to clicking the block.
clear	cl	clear *index* removes a breakpoint from the referenced block.
continue	c	continue resumes execution until reaching the next breakpoint or the end of the simulation. After you enter the continue command, it can be repeated by pressing the Enter key.
disp	d	disp *index* causes Simulink to display a block report for the specified block each time the simulation pauses for a breakpoint.
help	h	Display debugger help information.
ishow	i	Toggle display of integration information.
minor	m	Toggle display of minor integration step information.
nanbreak	na	Toggle breakpoint for undefined numeric data.
next	n	Proceed to the start of the next time step.
probe	p	probe *index* displays a block report for the referenced block. probe without an argument initiates interactive probing.
quit	q	Stop the simulation and quit debugging.
run	r	Resume simulation and ignore breakpoints.
slist	sli	Display a sorted list of all executable blocks in a model.
states	state	Display the current values of all model states.
status	stat	Display a debugger status report.
step	s	Single step to the next block.
stop	sto	Stop the simulation, and quit debugging.
systems	sys	List the nonvirtual systems.
tbreak	tb	tbreak *time* sets or clears a time breakpoint at the referenced time. tbreak with no argument sets or clears a time breakpoint at the current time.

TABLE 12.3: Debugger Command Summary (Cont)

Command	Short Form	Usage
trace	tr	trace *index* activates tracing for the referenced block.
undisp	und	undisp *index* cancels the disp command for the referenced block.
untrace	unt	untrace *index* cancels tracing for the referenced block.
xbreak	x	Toggles the stepsize limiting event breakpoint.
zcbreak	zcb	Toggles the zero-crossing event breakpoint.
zclist	zcl	Displays a list of blocks that may be associated with zero crossings.

12.5 SUMMARY

In this chapter, we discussed debugging Simulink models. We discussed techniques for locating model-building and implementation problems. We also discussed two aids for debugging Simulink models: floating scope and display blocks and the Simulink debugger.

13

Numerical Issues

In this chapter, we will discuss four numerical issues you should consider when building Simulink models. First, we will consider choosing the best differential equation solver for a particular model, which can frequently improve the speed and accuracy of a simulation. Next, we will explain algebraic loops, how Simulink deals with them, and how you can eliminate them if necessary. We will also discuss multiple sample time issues and zero-crossing issues.

13.1 INTRODUCTION

The Simulink block diagram metaphor for programming frees you from many of the details of writing a computer program to model a dynamical system. However, when you build a Simulink model, you are programming, and, as with all programming tasks, there are numerical issues you should keep in mind. We will discuss four of these issues in this chapter: choosing a differential equation solver, dealing with algebraic loops, multiple sample times, and zero-crossing detection.

13.2 CHOOSING A SOLVER

Simulink provides several differential equation solvers. The majority of the solvers are the result of recent numerical integration research and are among the fastest and most accurate methods available. Detailed descriptions of the algorithms are available in the paper by Shampine [4], available from The MathWorks.

It is generally best to use the variable-step solvers, as they continuously adjust the integration step size to maximize efficiency, while maintaining a specified accuracy. The Simulink variable-step solvers can completely decouple the integration step size and the interval between output points, so it is not necessary to limit the step size to get a smooth plot or to produce an output trajectory with a predetermined fixed-step size. The available solvers are listed in Table 4.5, repeated here for convenience in Table 13.1.

Solver Selection Considerations. There is no universal "best" differential equation solver. Choosing the best solver for a particular system requires an understanding of the system dynamics. Let's look briefly at the characteristics of each of the methods, and mention systems for which each is probably the best choice.

TABLE 13.1: Simulink Solvers

Solver	Characteristics
ODE45	Excellent general-purpose one-step solver. Based on the Dormand–Prince fourth-fifth order Runge–Kutta pair. ODE45 is the default solver and is usually a good first choice.
ODE23	Uses the Bogacki–Shampine second-third order Runge–Kutta pair. Sometimes works better than ODE45 in the presence of mild stiffness. Generally requires a smaller step size than ODE45 to get the same accuracy.
ODE113	Variable-order Adams–Bashforth–Moulton solver. Because ODE113 uses the solutions at several previous time points to compute the solution at the current time point, it may produce the same accuracy as ODE45 or ODE23 with fewer derivative evaluations, and thus perform much faster. Not suitable for systems with discontinuities.
ODE15S	Variable-order multistep solver for stiff systems. Based on recent research using numerical difference formulas. If a simulation runs extremely slowly using ODE45, try ODE15S.
ODE23S	Fixed-order one-step solver for stiff systems. Because ODE23S is a one-step method, it is sometimes faster than ODE15S. If a system appears to be stiff, it is a good idea to try various stiff solvers to determine which one performs best.
ODE23T	One-step solver for stiff systems, based on trapezoidal integration. ODE23T and ODE23TB are variants of the same method. ODE23T is somewhat faster but also less stable.
ODE23TB	Variant of ODE23TB using backward differentiation formulas for error estimation. This version is more stable than ODE23T but is also somewhat slower.
Discrete	Special solver for systems that contain no continuous states.
ODE5	Fixed-step version of ODE45.
ODE4	Classic fourth-order Runge–Kutta formulas using a fixed step size.
ODE3	Fixed-step version of ODE23.
ODE2	Fixed-step second-order Runge–Kutta method, also known as Heun's method.
ODE1	Euler's method using a fixed step size.

ODE45 and ODE23 are Runge–Kutta methods. These methods approximate the solution function (in the case of Simulink, the solution function is the state trajectory of the Simulink model) by numerically approximating a Taylor series of a fixed number of terms, the order being defined as the highest derivative in the series. The principal error in a Taylor series approximation of a function is known as *truncation error* and is due to the truncation of the Taylor series to a finite number of terms. ODE45 and ODE23 estimate the truncation error by computing the value of the state variables at the end of an integration step using two Taylor series approximations of different orders (4 and 5 or 2 and 3). The difference in the two computed values is a reliable indicator of the total truncation error. If the error is too large, the integration step size is reduced, and the integration step is repeated. If the error is too small (more accuracy than needed), the step size is increased for the next integration step. An important characteristic of algorithms such as ODE45 and ODE23 is that they select the intermediate points in the integration step such that both Taylor series approximations use the same derivative function evaluations.

ODE113 is a variable-order Adams method (namely, an Adams–Bashforth–Moulton method), a multistep predictor-corrector algorithm. The predictor step approximates the derivative function as a polynomial of degree $n - 1$, where n is the order of the method. The coefficients of the predictor polynomial are computed using the previous $n - 1$ solution points and the derivatives at the points. A trial next solution point is computed by extrapolation. Next, a corrector polynomial is fit through the previous n points and the newly computed trial solution point, and this polynomial is evaluated to recompute the trial solution point. The corrector portion of the algorithm can be repeated to refine the solution point. The difference between the predictor solution and the corrector solution is a measure of the integration error and is used to adjust the integration step size. ODE113 also adjusts the degree of the approximating polynomials to balance accuracy and efficiency. Multistep methods such as ODE113 tend to work very well for systems that are smooth. They don't work well for systems with discontinuities, because the polynomial approximation assumes a smooth function.

ODE15S is a variable-order multistep algorithm specifically designed to work well with stiff systems. A stiff system is one that has both very fast dynamics and very slow dynamics (widely separated eigenvalues). For example, many chemical processes involve sharp startup transients followed by relatively slow quasi-steady-state behavior. Numerical stiffness in mechanical systems can arise from a variety of factors, including physical stiffness (such as an assembly of physically stiff and physically limber components) or the use of very fast actuators to control slow dynamics. Special techniques, such as ODE15S, are required to accurately model such systems. These algorithms contain extra logic to detect transitions in a system's dynamics. The extra computational work expended in adapting to rapidly changing dynamics makes the stiff solvers inefficient for systems that are not stiff.

There are three fixed-order one-step stiff system solvers. Because these methods have no order adjustment logic, they are sometimes faster than ODE15S. ODE23S is based on the Rosenbrock formulas. ODE23T and ODE23TB are one-step stiff system solvers based on trapezoidal integration, differing only in the method of estimating truncation error. ODE23T is somewhat faster than ODE23TB but also less stable.

The discrete solver is a special method that is applicable only to systems that have no continuous states. Although all of the Simulink solvers are suitable for such systems, the discrete solver is the fastest choice for these systems.

The Solver options section contains four fields to control integration step size adjustment for the variable step size integrators. Two fields, **Max step size** and **Initial step size**, permit you to reduce the likelihood of the solver missing important system behavior. To allow Simulink to use its default values for these parameters, enter auto in the respective fields. The other two fields allow you to set the absolute and relative tolerances used in the step size adjustment logic.

The default **Max step size** is

$$h_{max} = \frac{t_{stop} - t_{start}}{50}$$

which is generally satisfactory. There may be certain situations in which it is desirable to enter a fixed value for **Max step size**. For example, if the duration of the simulation is extremely long, the default maximum step size may be too large to guarantee that no important behavior will be missed. If the system is known to be periodic, the performance of the step-size adjustment logic may be slightly improved by limiting the maximum step size to a fraction (The MathWorks suggests 1/4) of the period. It is not advisable to set the maximum step size so as to limit the spacing between points in the output trajectory, as it is much more economical computationally to use the Output options section of the solver page for that purpose.

If **Initial step size** is set to auto, Simulink will compute the initial step size based on the state derivatives at the start of the simulation. If the system dynamics are believed to contain a sharp transient soon after t_{start}, set **Initial step size** to a value small enough to permit the solver to detect the transient. In most other situations, it is best to allow the built-in logic to compute the initial step size.

Relative tolerance and **Absolute tolerance** are used to compute the allowable value of integration error estimate (e_i) for each state (x_i) according to the formula

$$e_i \leq \max(tol_{Rel} \mid x_i \mid, tol_{Abs})$$

where tol_{Rel} and tol_{Abs} are **Relative tolerance** and **Absolute tolerance**, respectively. The two tolerances for a hypothetical state are depicted graphically in Figure 13.1. In the region in which the magnitude of the state is large, tol_{Rel} determines the error bound. In the region where the magnitude of the state is small, tol_{Abs} determines the error bound. If the integration error estimate for any state exceeds its limit, the integration step size is reduced. If the error limits for every state exceed the

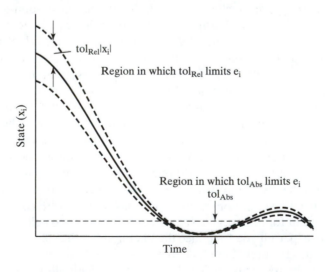

FIGURE 13.1: Solver error tolerance computation

estimates by some value that depends on the particular solver used, the integration step size for the subsequent step is increased.

The fixed-step-size solvers are all fixed-order one-step methods. These methods might be preferable if the system dynamics are sufficiently well understood that a nearly optimal integration step size is known. In such a situation, the elimination of step size and order adjustment might significantly speed up the simulation. It is not advisable to use a fixed-step-size integrator to force Simulink to produce an output trajectory with a fixed spacing between points, as the Output options section of the **Solver** page allows you to do that much more economically. If a fixed-step-size integrator is chosen, there is a single field, Fixed step size, to enter the step size. This field may contain a value for step size or may be set to auto to allow Simulink to automatically choose the fixed-step size.

EXAMPLE 13.1   Stiff system

A stiff system is a system that has both fast dynamics and slow dynamics. Typically, the primary interest is in the slow dynamics, but ignoring the fast dynamics can cause the simulation to produce incorrect results. Consider the unforced second-order system

$$\ddot{x} + 100\dot{x} + 0.9999x = 0$$

Assume that the system starts at rest with $x = 1$. Taking the Laplace transform,

$$s^2 X(s) - s\,x(0) - \dot{x}(0) + 100(s\,X(s) - x(0)) + 0.9999X(s) = 0$$

Substituting $x(0) = 1$, $\dot{x}(0) = 0$, and solving for $X(s)$,

$$X(s) = \frac{s + 100}{s^2 + 100s + 0.9999}$$

Taking the inverse Laplace transform, we get the time response

$$x(t) = -0.0001e^{-99.99t} + 1.0001e^{-0.01t}$$

The response of this system has two components. The first component starts at a very small magnitude (-0.0001) and decays rapidly. The second component starts at a magnitude 10,000 times as large and decays 10,000 times as slowly. So, the slow response dominates the behavior of the system.

Now, let's see what happens when we model this system. Shown in Figure 13.2 is a Simulink model of the system. Set the **Initial condition** of the velocity Integrator to 0, and

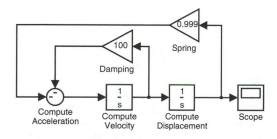

FIGURE 13.2: Simulink model of stiff second-order system

set the **Initial condition** of the displacement Integrator to 1. Set **Start time** to 0, and set **Stop time** to 500. Select (check) **Simulation:Parameters** dialog box **Workspace I/O** page field **Save to workspace, Time**, and leave **States** and **Output** unselected.

Next, let's experiment with the different solvers to see the effect of the stiffness. First, run the simulation using ODE15S. The simulation runs in a few seconds. The tout (produced as a result of selecting **Save to workspace, Time**) and xout (produced by the To Workspace block) vectors sent to the MATLAB workspace each have about 100 elements. Shown in Figure 13.3 are the simulation results plotted using the MATLAB plot command.

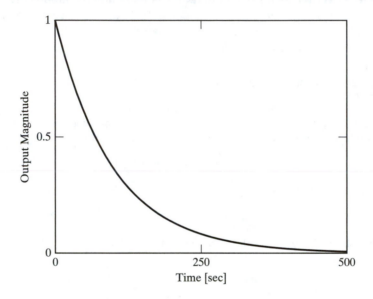

FIGURE 13.3: Stiff system results using ODE15S

Running the simulation using ODE45 takes a much longer time but produces a plot that appears identical to that in Figure 13.3. This time, however, the vectors tout and xout have approximately 15,000 elements each. If the output (xout) and time (tout) are sent to the MATLAB workspace, the following MATLAB statements will produce a plot showing the change in the trajectory between successive points from point 14,000 to 14,100:

```
t=tout(14000:14100);

x=xout(14001:14101)-xout(14000:14100);

plot(t,x)
```

The plot appears in Figure 13.4. Inspecting the plot, you can see that there is a high-frequency component to the output trajectory resulting from the fast dynamics. This component is of very small magnitude (10^{-6}).

Finally, recall that we said that ignoring the fast dynamics can produce incorrect results. The stiff solver required about 100 time points, or on average 5 sec between time points. If we run the simulation using the fixed-step solver ODE4 and a 5 sec step size, the simulation diverges, causing Simulink to issue an error message and stop.

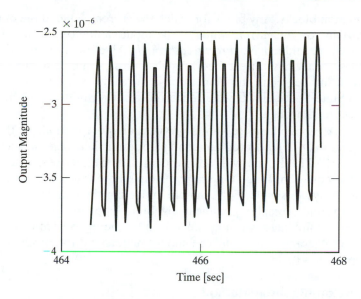

FIGURE 13.4: Oscillation of stiff system results using ODE45

13.3 ALGEBRAIC LOOPS

Algebraic loops are a programming issue that sometimes requires special care in Simulink modeling. An algebraic loop is a condition in which the output of a block drives the input of the same block. Consider the Simulink model in Figure 13.5. From the model, we can compute

$$\dot{x} = u - k_2 k_3 \dot{x} - k_1 k_3 x \tag{13.1}$$

So $\dot{x}$ is a function of x, u, and $\dot{x}$. For each integration step, Simulink must solve the algebraic equation for $\dot{x}$. Simulink cannot solve for $\dot{x}$ symbolically, so it uses an iterative numerical technique to solve the algebraic equation. This iterative procedure takes time, and in some cases, Simulink fails to arrive at a solution altogether.

   In this simple example, the algebraic loop is formed by Gain blocks k2 and k3 and Sum blocks Sum and Sum1. However, algebraic loops are not restricted to Gain

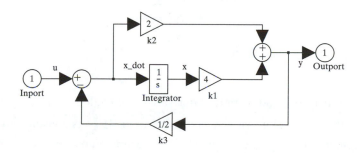

FIGURE 13.5: First-order system with an algebraic loop

and Sum blocks. Any block for which the current value of the output depends, even partially, on the current value of the input can be part of an algebraic loop. Such blocks are said to have *direct feedthrough*. All of the blocks in the Math Operations block library have direct feedthrough. Transfer Function and Zero-Pole blocks for which the degree of the numerator is the same as the degree of the denominator, and State-Space blocks with nonzero direct transmittance (D) matrices also exhibit direct feedthrough.

Algebraic loops are not a problem unique to Simulink. If you attempt to use Equation (13.1) in a program in an algorithmic language such as FORTRAN, C, or even MATLAB, you will have to deal with the fact that $\dot{x}$ appears on both sides of the equation. While the scalar example here is easily dealt with, the situation is not always so simple. For example, modeling the dynamics of robot manipulators can produce complex nonlinear algebraic loops involving several state variables. The Simulink algebraic loop solver can frequently relieve you of the need to worry about algebraic loops, but you should understand the problem and how Simulink approaches it.

13.3.1 Newton–Raphson Method

Simulink attempts to solve algebraic loops using a robust implementation of the Newton–Raphson technique. To illustrate the basic idea of the technique, we will discuss the classic Newton–Raphson method for a scalar problem. For the scalar example above, the algebraic problem can be written

$$\dot{x} = f(x, \dot{x}, u, t)$$

where f is, in general, a nonlinear function. The Newton–Raphson method is an iterative process that attempts to solve the error function

$$\phi = \dot{x} - f(x, \dot{x}, u, t) = 0$$

by minimizing the quadratic

$$P = \phi^2$$

with respect to $\dot{x}$. At each iteration, compute (dropping the arguments of ϕ for convenience)

$$\Delta\dot{x} = -0.5\phi \left/ \frac{\partial\phi}{\partial\dot{x}} \right.$$

then update the estimate

$$\dot{x}_{new} = \dot{x}_{old} + \Delta\dot{x}$$

The function ϕ and its partial derivative are evaluated at the current estimate of $\dot{x}$, and the current known values of x, u, and t. If $\dfrac{\partial\phi}{\partial\dot{x}}$ is constant, that is, if $f(x, \dot{x}, u, t)$ is linear with respect to $\dot{x}$, the Newton–Raphson procedure will converge in exactly one iteration. On the other hand, if $f(x, \dot{x}, u, t)$ is nonlinear with respect to $\dot{x}$, the procedure may require many iterations and may fail to converge altogether.

13.3.2   Eliminating Algebraic Loops

Simulink will report the detection of an algebraic loop if the **Simulation:Parameters** Diagnostic choice for algebraic loops is set to warning or error. If an algebraic loop is detected, you have two options: leave the algebraic loop intact, or eliminate it. If the speed of execution of the model is acceptable, leaving the loop intact is likely the better choice. If the speed of execution is not adequate, you must eliminate the algebraic loop.

The most desirable method of eliminating an algebraic loop is to reformulate the model into an equivalent model that does not have an algebraic loop. For example, the model shown in Figure 13.6 has the same input–output behavior as the model in Figure 13.5. However, this model does not have an algebraic loop and will therefore execute faster.

It is not always convenient to reformulate a model such that there are no algebraic loops, as the equations of motion of some physical systems lead to algebraic loops. It is always possible in these situations to break an algebraic loop using a memory block. Shown in Figure 13.7 is the model of Figure 13.5 modified such that the algebraic loop is broken using a Memory block. While this approach can always be used to break an algebraic loop, it is not always satisfactory, because the delay introduced by a Memory block can degrade the accuracy of the simulation. As we will see in Example 13.2, this approach can change the behavior of the Simulink model so that it no longer accurately represents the physical system.

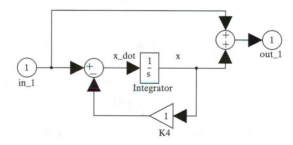

FIGURE 13.6: First-order system reformulated to eliminate the algebraic loop

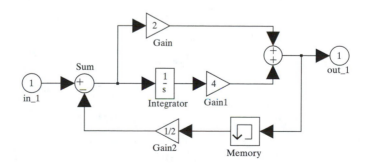

FIGURE 13.7: Breaking an algebraic loop with a memory block

A final method of eliminating an algebraic loop is to use a MATLAB function block to solve the algebraic problem directly. Although this approach requires that you write MATLAB code to replace some Simulink blocks, it is sometimes the best solution.

_____

EXAMPLE 13.2   Directly coupled two-cart system

Consider the system of two carts shown in Figure 13.8. Cart 1 moves on the ground, and Cart 2 moves relative to Cart 1. Variables x_1 and x_2 are the positions of the carts relative to inertial space. The angular displacements of the wheels are θ_1 and θ_2, and the total torque input to the wheels of Cart 1 is τ_1, and to the wheels of Cart 2 is τ_2. The wheel angular displacements are related to the cart absolute positions by

$$x_1 = -r_1\theta_1$$

$$x_2 = x_1 - r_1\theta_2$$

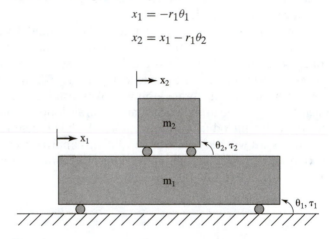

FIGURE 13.8: Directly coupled carts

The kinetic energy of the system in terms of the wheel angular velocities is

$$T = \frac{1}{2}m_1\left(-r_1\dot{\theta}_1\right)^2 + \frac{1}{2}m_2\left(-r_1\dot{\theta}_1 - r_2\dot{\theta}_2\right)^2$$

Using a Lagrangian approach, we solve for the wheel torques:

$$\tau_1 = m_1 r_1^2 \ddot{\theta}_1 + m_2 r_1^2 \ddot{\theta}_1 + m_2 r_1 r_2 \ddot{\theta}_2$$

$$\tau_2 = m_2 r_1 r_2 \ddot{\theta}_1 + m_2 r_2^2 \ddot{\theta}_2$$

Solving for the angular accelerations, we get the coupled differential equations

$$\ddot{\theta}_1 = K_{11}\left(\tau_1 - K_{21}\ddot{\theta}_2\right)$$

$$\ddot{\theta}_2 = K_{22}\left(\tau_1 - K_{12}\ddot{\theta}_1\right)$$

where

$$K_{11} = \frac{1}{(m_1 + m_2)\, r_1^2}$$

$$K_{21} = K_{12} = m_2 r_1 r_2$$

$$K_{22} = \frac{1}{m_2 r_2^2}$$

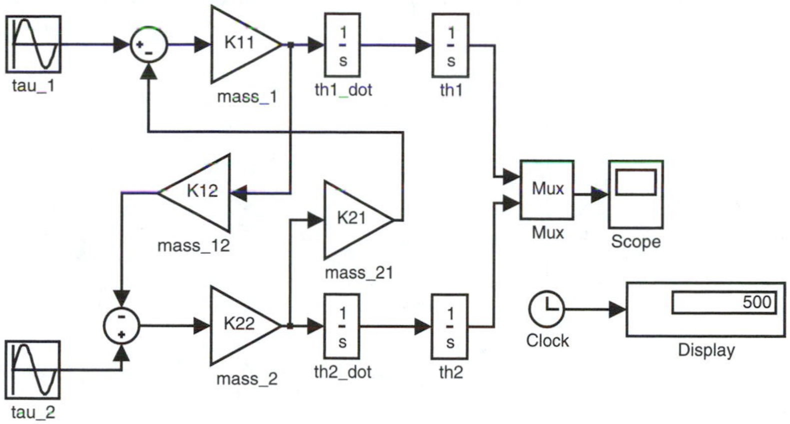

FIGURE 13.9: Simulink model of directly coupled carts

A Simulink model that implements the equations of motion is shown in Figure 13.9. We set the parameters using the M-file shown in Listing 13.1.

Listing 13.1: MATLAB script to initialize coefficients

```
% Set up block parameters for algebraic loop example
m1 = 10 ;
m2 = 5 ;
r1 = 2 ;
r2 = 1.5 ;
K11 = 1/((m1+m2)*r1^2) ;
K12 = m2*r1*r2 ;
K21 = K12 ;
K22 = 1/(m2*r2^2) ;
```

The input torques are

$$\tau_1 = \sin(0.1t)$$

$$\tau_2 = \sin(0.2t - 1)$$

Executing the simulation produces an algebraic loop warning in the MATLAB workspace. The output trajectories of the two carts for the first 500 sec are shown in Figure 13.10.

To illustrate the potential for simulation errors when an algebraic loop is broken using a Memory block, consider the revised model in Figure 13.11. The Memory block eliminates the algebraic loop and causes the simulation to execute significantly faster. Unfortunately, the results of the simulation, shown in Figure 13.12, are incorrect. If the simulation is run with the **Max step size** on the **Simulation:Parameters Solver** page set to a sufficiently small value, correct results are produced. In this example, if the **Max step size** is set to 0.1, the results are nearly identical to the results shown in Figure 13.10. Of course, limiting **Max step size** slows the simulation, diminishing the benefit of the Memory block.

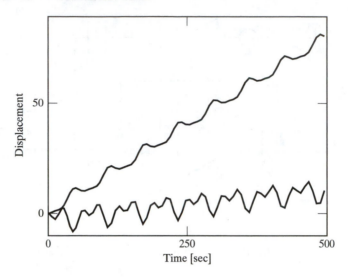

FIGURE 13.10: Directly coupled cart trajectories

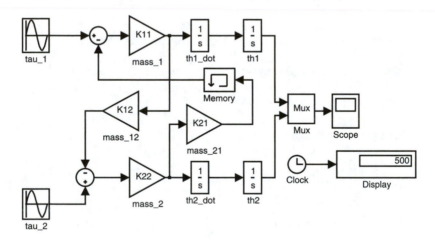

FIGURE 13.11: Directly coupled cart model with broken algebraic loop

13.4 MULTIPLE SAMPLE TIME ISSUES

A Simulink model with discrete states can have several different sample times. It is important, in the interest of computational efficiency, to choose sample times that are a relatively small multiple of a single sample time, which could be called the fundamental sample time. To illustrate the issues, suppose a discrete model has two sample times: 1/2 sec and 1/4 sec. The fundamental sample time will be 1/4 sec. Now suppose the sample times are 1/17 sec and 1/19 sec. In this case, the largest sample time that will produce samples every 1/17 sec and every 1/19 sec is $1/[(17)(19)]$, or approximately 0.003 sec. Adding more sample times in this situation has the potential to cause the simulation to execute extremely slowly.

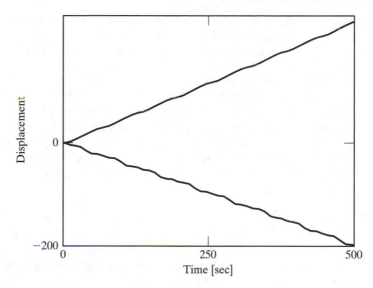

FIGURE 13.12: Trajectories using the directly coupled cart model with the algebraic loop broken

13.5 ZERO-CROSSING DETECTION

A number of Simulink blocks exhibit discontinuous output when the continuous input crosses some threshold. Among these blocks are the Relational Operator, Sign, and Relay. For example, the output of the Sign block changes from -1 to 1 as the input increases through zero. When Simulink is configured to use a variable-step-size solver, it will automatically adjust the step size so as to accurately locate the zero crossing. If the signal crosses zero infrequently, zero-crossing detection greatly improves the accuracy of the simulation and has no undesirable effects. On the other hand, if the signal crosses zero frequently, a phenomenon known as chatter can occur, causing the simulation to execute extremely slowly. See Section 12.3 for details on locating and correcting zero-crossing detection problems.

13.6 SUMMARY

In this chapter, we discussed two important numerical issues you should consider when building Simulink models. First, we discussed choosing the most appropriate solver, which can improve the speed and accuracy of a simulation. Then, we discussed algebraic loops, how Simulink deals with them, and how you can eliminate them.

REFERENCES

[1] Hartley, Tom T., Beale, Guy O., and Chicatelli, Stephen P., *Digital Simulation of Dynamic Systems: A Control Theory Approach* (Englewood Cliffs, N.J.: Prentice Hall, 1994), 190–238. In addition to covering the most important differential equation solution techniques, this book provides a detailed discussion of solutions to stiff systems.

[2] Kincaid, David, and Cheney, Ward, *Numerical Analysis* (Pacific Grove, Calif.: Brooks/Cole Publishing Co., 1991), 508–514. This is an excellent, and rigorous, text covering many important topics in numerical analysis.

[3] Mathews, John H., *Numerical Methods for Mathematics, Science, and Engineering* (Englewood Cliffs, N.J.: Prentice Hall, 1992). This is a fine numerical methods reference for scientists and engineers. In particular, Chapter 4 on interpolation and polynomial approximation and Chapter 9 on differential equation solution will assist you in understanding Simulink solvers. Chapter 2 provides a rigorous treatment of Newton–Raphson methods. Many algorithms are provided, and MATLAB code for the algorithms is available from The MathWorks Web site (www.mathworks.com).

[4] Shampine, Lawrence F., and Reichelt, Mark W., "The MATLAB ODE Suite" (The MathWorks, Inc., Natick, Mass.: 1996). This technical paper is available directly from The MathWorks. It provides a detailed discussion of the MATLAB differential equation solvers that are available from within Simulink. It also provides an extensive list of references.

14

Introduction to Stateflow

In this chapter we will discuss Stateflow. We'll begin with an introduction to finite state machines and the Statechart representation of finite state machines. Then we will discuss the basics of building Stateflow charts and the interactions between Stateflow and Simulink models.

14.1 INTRODUCTION

Stateflow is an extension to Simulink that provides a powerful environment with which to add finite state machines to Simulink models. Stateflow is built around the Statechart formalism [3,4–7] which has gained considerable popularity over the last several years. While it is, of course, possible to model finite state machine logic using standard Simulink logic blocks, Stateflow adds the ability to model finite state machine subsystems using a notation that is specifically designed for this purpose.

The objective of this chapter is to introduce Stateflow and illustrate how Stateflow blocks are built and incorporated into Simulink models. We will begin with an overview of finite state machines and state transition diagrams. Next, we will introduce Statecharts. Then, we will illustrate building Stateflow charts and discuss the various ways Stateflow charts can interact with Simulink models. Stateflow is powerful and complex, and therefore, it is only possible to present a brief introduction here. However, once you master the information provided here, you should be able to build simple Stateflow charts and progress to more complicated systems.

14.2 FINITE STATE MACHINES

Many dynamical systems have subsystems that are always in one of a finite set of configurations. Such subsystems are called *finite state machines*, and each possible configuration is a called a *state*. A state can be thought of as a vector of a finite number of elements, much like the continuous and discrete state vectors that we discussed in Chapter 8. A typical finite state machine subsystem is the switching logic that controls an aircraft landing gear. This logic includes the landing gear control handle in the cockpit and various sensors and switches designed to prevent premature retraction of the landing gear and to ensure that the landing gear is lowered when the control handle is placed in the down position.

TABLE 14.1: Light Switch Truth Table

Switch State	Light Status
Up	On
Down	Off

TABLE 14.2: Two-switch Truth Table

Switch A	Switch B	Light Status
Up	Up	Off
Down	Up	On
Down	Down	Off
Up	Down	On

There are many means with which to represent finite state machines. Perhaps the simplest is the truth table. Suppose we wish to model a light switch. We could represent the light switch system using the truth table shown in Table 14.1. This system has two states: Up and Down.

Now, suppose we wish to model a room system in which two switches control a single light. Each switch has two possible values (Up, Down), so there are a total of four possible configurations of the system (four states), as shown in Table 14.2.

The truth table specifies the possible configurations of the system, but it doesn't indicate any restrictions on transitions from one state to another. Often, it is not possible to transition from each possible state to every other possible state. For example, the two-light-switch system may allow only one switch to change at a time. The need to illustrate the allowable states and allowable transitions can be met using state transition diagrams.

14.3 STATE TRANSITION DIAGRAMS

A state transition diagram graphically depicts states and transitions between states. Figure 14.1 is a state transition diagram for the single-light-switch system. There are two states (On, Off) and a transition from each state to the other. The state transition diagram contains three types of objects. First, there are states, represented by the circles (sometimes called bubbles). The lines connecting the states are the permissible transitions. The label on each transition represents the event that causes the transition. In this simple system, there is a one-to-one relationship between the transitions and the events, but this is not, in general, necessary.

Figure 14.2 is a state transition diagram for the two-switch room. Here the transitions represent the changing of position of a single switch. For example, LU represents switching the left switch up. The state transition diagram depicts the four possible states, the allowable transitions among the states, and the events that cause transitions. Notice that in this state transition diagram, there are different outcomes of each event, depending on the system state when the event occurs.

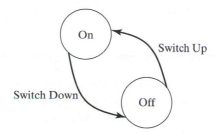

FIGURE 14.1: Single-switch system state transition diagram

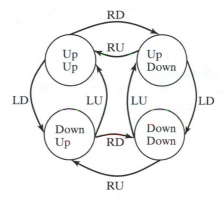

FIGURE 14.2: Two-switch state transition diagram

The combination of truth table and state transition diagram specifies the allowable states, the results of being in each state (light on or off), and the allowable transitions between states. In many cases, this combination is adequate. However, as system complexity grows, this combination can become unwieldy due to the large number of states.

14.4 Stateflow CHARTS

A Stateflow chart is an enhanced state transition diagram based on Statechart notation [3,4–7]. The enhancements include the addition of hierarchy to provide various levels of abstraction and parallelism to facilitate modeling systems with independent finite state machine subsystems. Another enhancement that Stateflow provides is the association of actions with transitions and states. The association of actions with transitions and states eliminates the need for truth tables to supplement the state transition diagrams and also allows for more flexibility in specification. For example, it is possible to indicate that an action (such as turning the light on or off) occurs as a result of a transition, as a result of entering a particular state (which may be reachable via more than one transition), or upon leaving a state. Stateflow charts also provide for branching via compound transitions that can include sequential events and decision points.

Stateflow transition labels have four parts, each of which is optional. The format is

`event[condition]condition action/transition action`

Here, *event* has the same meaning as it does for a state transition diagram. The *condition* is a logical expression that must evaluate to true in order for the transition to take place. Thus, if *event* occurs and *condition* is false, the transition does not occur. *condition action* is any behavior that is to occur if *condition* is true, whether or not the transition takes place. *transition action* is any behavior that is to occur as a by-product of the transition. If *event* is not present in the label, but *condition* is present, the transition is triggered when *condition* becomes true. Stateflow provides for compound transitions, to be discussed shortly. In the case of a compound transition, it is possible for the condition action to take place but not the transition action.

In a case of a simple finite state machine, the Stateflow chart appears similar to a state transition diagram. For example, shown in Figure 14.3 is a Stateflow representation of the single-switch system. Notice that the circles are replaced with rounded rectangles, and also notice that the transition label consists of two parts: an event and an action in the format

`event/event action`

In this case, the event is changing the position of the switch, and the actions are turning the light on or off. Also note that there is an additional transition that originates in a small filled circle. This transition is called a *default transition*, and it serves to specify the initial state of the system.

A Stateflow chart representing the two-switch system is shown in Figure 14.4. Again, note the use of the default transition to specify the initial state of the system. However, in this example, conditions (`[L==1]`, etc.) are specified rather than events. Thus, a transition takes place when the specified condition becomes true. Also, the actions (`Light=0 or 1`) are associated with the four states rather than the transitions. Therefore, each state label includes the state name (for example, UpUP) and the action associated with the state. Because the actions are associated with the states rather than the transitions, the transitions are labeled with only the conditions.

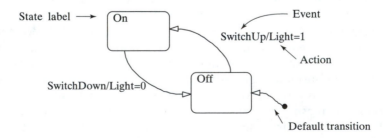

FIGURE 14.3: One-switch Stateflow chart

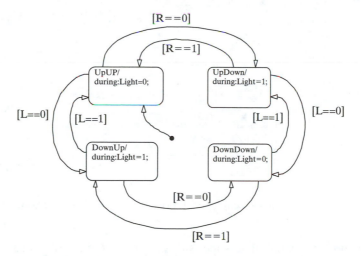

FIGURE 14.4: Two-switch Stateflow chart

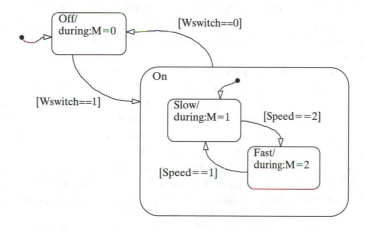

FIGURE 14.5: Windshield wiper system

14.4.1 Stateflow Chart Hierarchy

An important aspect of Stateflow charts is hierarchy. In an hierarchical Stateflow chart, states can be contained within other states. As a simple example, consider a windshield wiper system. The system has two states that can be termed superstates: On ($M \neq 0$) and Off ($M = 0$), where M is the motor speed command. The On superstate has two substates: Slow ($M = 1$) and Fast ($M = 2$). A Stateflow chart of the windshield wiper system is shown in Figure 14.5. Notice that the default transition for the system is to state Off. Once the system transitions to the On superstate, that state starts in the Slow substate.

A second form of decomposition is termed AND or parallel decomposition, which denotes orthogonal or parallel states. To illustrate AND decomposition, consider Figure 14.6, which depicts a partial model of a car electrical system that

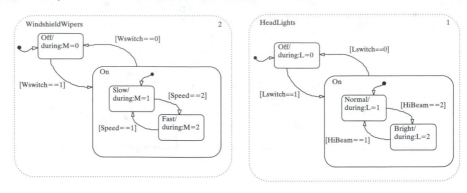

FIGURE 14.6: Car electrical system: AND decomposition

includes the windshield wiper subsystem and the headlight subsystem. The dashed borders of the windshield wiper and headlight subsystems indicate the parallel construction and indicate that the two subsystems are independent. Thus, it is possible to depict the state of the headlights and the state of the windshield wipers in the same diagram. The overall state of the electrical system includes the windshield wipers and headlights, and the overall state will determine, for example, the total electrical load.

Another useful element of Stateflow charts is the history junction. A history junction is represented by a circle containing the letter H. If a superstate contains a history junction, the state that was active the last time the superstate was entered will be the default state the next time the superstate is entered. The first time the superstate is entered, the default transition will be executed, but the default transition will thereafter be overruled by the history junction.

To illustrate, consider the revised headlight subsystem shown in Figure 14.7. The On superstate contains a history junction. The first time the On superstate is entered, the default transition to Normal will take place. Now suppose the lights are

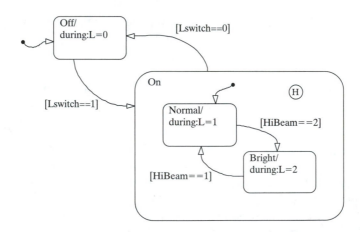

FIGURE 14.7: Headlight system with history junction

switched to Bright and then turned off. The next time they are turned on, they will start Bright.

14.4.2   Compound Transitions

Stateflow notation provides a powerful mechanism to simplify charts via *compound transitions*. A compound transition contains connective junctions that permit decision making and branching. A connective junction is represented by a small circle. The two basic types of compound transitions are *forks* and *joints*. A fork has a connective junction that splits a transition into two or more alternatives. A joint is a connective junction that merges two or more transitions. We will briefly discuss each type.

Consider Figure 14.8. This figure illustrates a fork connective junction. If the system is in State A and event E1 occurs, the system will transition to State B if variable A is 1 and to State C if A is 0. If neither condition is true (A is neither 1 nor 0), neither transition will occur, and State A will remain active. Note that the connective junction is not an intermediate state, rather it is the equivalent of a logical AND test.

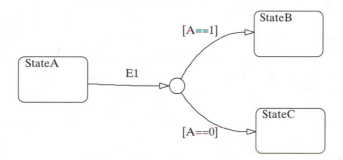

FIGURE 14.8: Fork compound transition

Next, to illustrate a joint, consider Figure 14.9. In this example, if either State A or State B is active when event E1 occurs, the system will transition to State C. Because there is no label on the transition segment from the connective junction to State C, that part of the compound transition is unconditional.

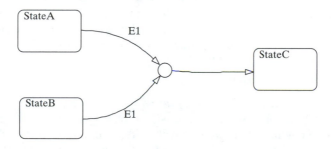

FIGURE 14.9: Joint compound transition

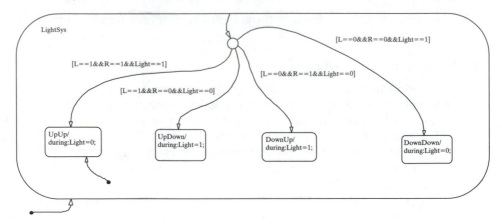

FIGURE 14.10: Two-switch system using inner transition

Compound transitions can greatly simplify a Stateflow chart, particularly in systems that involve a large amount of conditional logic.

14.4.3 Inner Transitions

Inner transitions can often simplify a Stateflow chart. An inner transition is a transition from a superstate to a connective junction or substate in the same superstate. An inner transition is represented as a transition line drawn from the border of the superstate to a connective junction or substate in the same superstate. The inner transition is equivalent to a transition from every substate in the superstate to the inner transition destination.

To illustrate an inner transition, consider Figure 14.10. This Stateflow chart is equivalent to the two-switch Stateflow chart shown in Figure 14.4. The first time the chart updates, the default transition into superstate LightSys takes place, followed immediately by the default transition to substate UpUp. Thereafter, each time the chart updates, the unconditional inner transition to the connective junction takes place, and if any of the four transition conditions leaving the connective junction is valid, the corresponding transition is completed. Otherwise, the transition is not completed, and the current substate remains active. Observe that if the test for the value of Light were not present in each condition, the active substate would be exited and reentered each time the chart is updated. Therefore, without the test for the value of Light, the behavior of this chart would be quite different from the behavior of the two-switch chart shown in Figure 14.4.

14.5 Stateflow QUICK START

To get started with Stateflow, we will build a simple Simulink model that uses a Stateflow block to set an output signal based on events that occur in the Simulink model. Specifically, the system will produce an output of 1.0 when the input is in the range $0 \le u \le 0.5$ and the value of u is increasing, and it will produce an output of 0.0 otherwise.

First, open a new model window and begin building the model as shown. Configure the Hit Crossing block labeled Cross zero rising. Set the block dialog box field **Hit crossing offset** to 0 and **Hit crossing direction** to rising. Configure Hit Crossing block Cross 1/2 **Hit crossing offset** to 0.5 and **Hit crossing direction** to either.

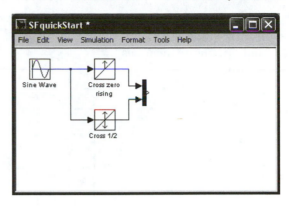

Drag a Chart block from the Stateflow block library. Position it as shown.

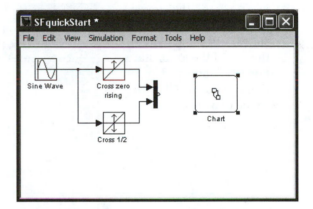

Double-click the Chart block, opening the Stateflow window. Here, the window was resized to fit the page. Notice that there are four icons on the left side of the window. These are the drawing icons used to build the Stateflow chart. Below the drawing icons is a zoom control that allows you to enlarge a portion of the chart.

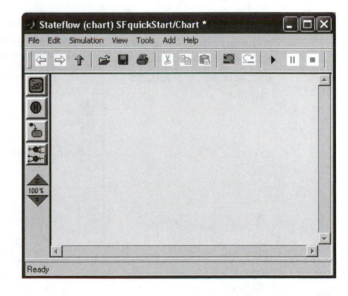

Drag two state boxes from the drawing palette, and position them as shown.

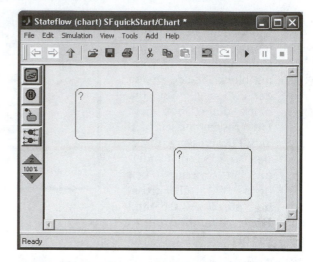

Next, draw the transitions between the two states. Click and hold on the top of the lower state box as shown. Notice that the cursor takes the shape of a cross.

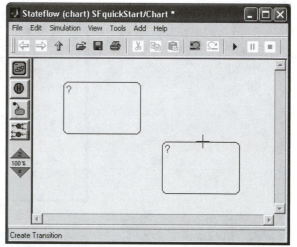

Drag to the edge of the other state box, and release the mouse button.

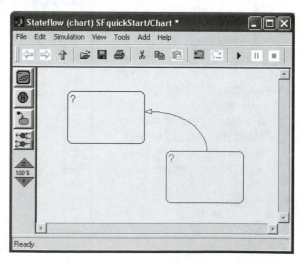

Draw the transition from
the upper state box to the
lower state box similarly.

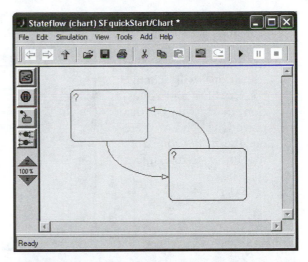

To install the default
transition, click the
default transition icon,
and drag to the edge of
the upper state box as
shown.

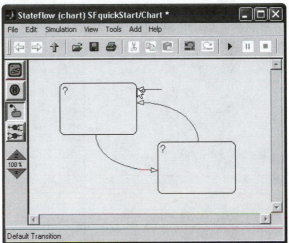

Release the mouse but-
ton.

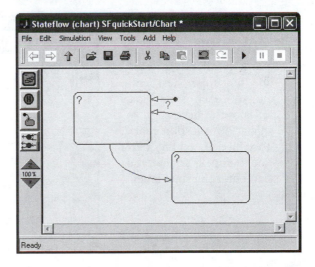

Next, we will label the transitions and states. To label a transition, click the transition once. The transition will change color to blue, and a question mark will be displayed as shown here on the transition from the lower to the upper state box.

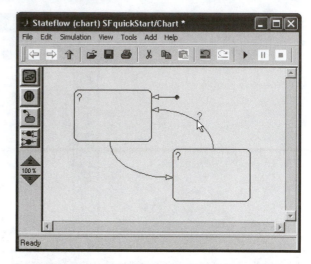

Now, click the question mark, and an editing cursor appears.

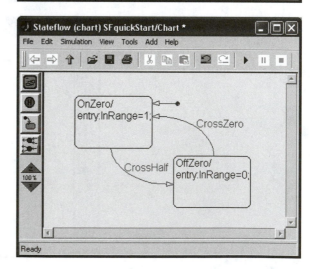

Label the transitions and states as shown. Note that because the question mark is already present in the upper left corner of each state box, it is only necessary to click the question mark to cause an editing cursor to appear so that you can enter the state label.

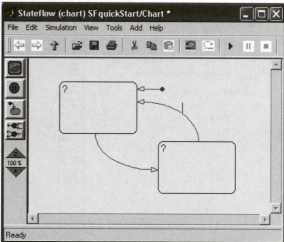

Next, define the two events. From the Stateflow chart window menu bar, select **Add:Event:Input from Simulink**. Configure the dialog box as shown here. Add a second event CrossHalf, and set field **Trigger** to Either edge. Note that the Index of the first event is 1, and the second event is 2.

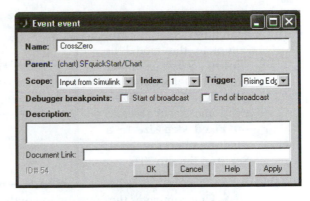

Now, define the output by selecting **Add:Data:Output to Simulink** from the Stateflow chart window menu bar. Configure the dialog box as shown and then click **OK**.

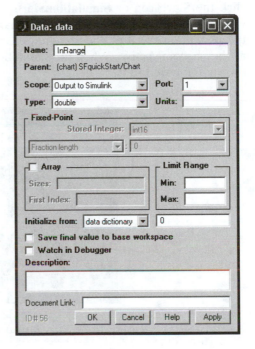

Now, notice that the Stateflow block has a trigger port and an output port. The trigger port expects a vector signal of events in the order defined in the Stateflow chart. The output signal will be the variable InRange defined in the Stateflow chart.

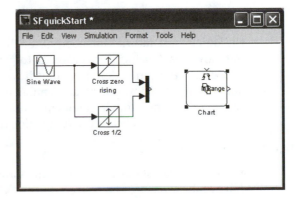

Rotate and resize the Stateflow block, and complete the model as shown. Select **Simulation:Parameters** and set the solver **Type** to Fixed step and Discrete and set **Fixed step size** to a small number, say 0.004. Set **Stop time** to 20.0.

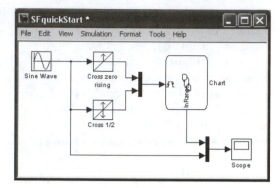

That completes the construction of the model. Open the Scope block, and then click the Start icon or **Simulation:Start**. There will be a delay of several seconds while Simulink compiles the Stateflow chart, and then the simulation will begin. As the simulation executes, the Stateflow chart transition and states will change colors as they become active. When the simulation stops, the Scope display should be as shown in Figure 14.11.

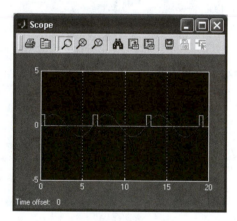

FIGURE 14.11: Stateflow quick-start model output

14.6 DRAWING Stateflow CHARTS

The quick-start tutorial included all of the basic procedures you'll need to create Stateflow charts and interface them to Simulink models. In this section, we will provide a little more detail on the drawing commands. As we've seen, drawing and editing Stateflow charts are similar to drawing and editing Simulink models. There are only five drawing elements: four on the palette (states, default transitions, history junctions, and connective junctions) and transitions.

14.6.1 State Boxes

All state boxes (states and superstates) are drawn using the state box icon on the chart window palette. To resize a state box, select it and drag an edge or corner.

To create a superstate, drag a state box to the drawing area and then enlarge it. You can either enlarge it to surround substates already drawn, or draw the superstate box first and then draw the substates within it.

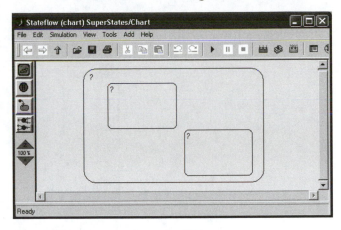

To delete a state, select it and press the Delete key.

14.6.2 Transitions

You've already learned to draw transitions by dragging from an edge of a source state to an edge of a destination state. You can move either the source or destination of a transition, and you can also reshape a transition.

To move either end of a transition, place the cursor over the end you wish to move. The cursor will change to a small circle.

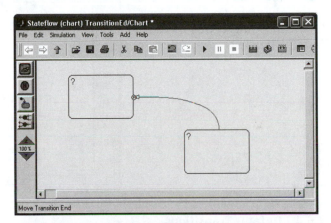

Drag the end of the transition to the new location. It can be on the same state (or connector) or a different one.

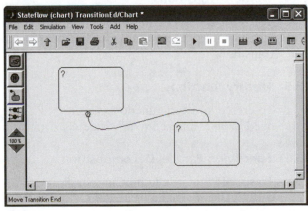

To reshape a transition, drag a point on the transition.

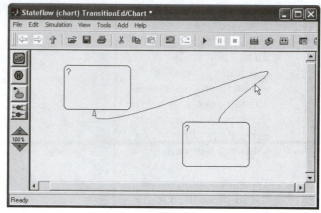

To move a transition label, select the label and drag it to the desired position.

14.6.3 Connective Junctions

To draw a fork or joint transition, drag a connective junction from the palette, and draw the transitions. Here, we illustrate with a fork transition.

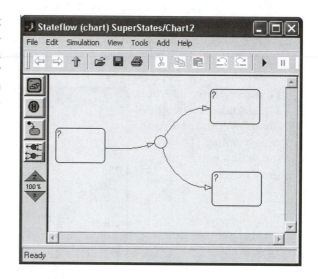

14.6.4 Default Transitions

We drew a default transition in the quick-start example. You can edit the path, location, and destination of a default transition in a manner identical to editing transitions.

14.6.5 History Junctions

To create a history junction, drag it into a superstate. A history junction does not need to be connected to any state via a transition.

14.6.6 Specifying Parallel Decomposition

To specify that a state box has parallel decomposition, right-click the box and then choose **Decomposition:Parallel (AND)** from the pop-up dialog box. The border of the box will become dashed. You can reverse the process by right-clicking the state box and choosing **Decomposition:Exclusive (OR)**.

14.7 LABELING STATES AND TRANSITIONS

Stateflow provides a rich set of capabilities for labeling states and transitions. In this section, we will provide a brief overview of the available options.

14.7.1 State Labels

The general format of a state label is as follows:

state name/

entry:*entry action*;

during:*during action*;

exit:*exit action*;

The *state name*/ is an optional name of the state. State names, if used, should be unique. There are three action conditions (entry, during, exit), each of which is optional, and each of which may be associated with one or more actions if present. Notice that the condition is followed by a colon (:) that serves as a separator between the condition and action. The action can be an event or an assignment statement. More than one action may be associated with each condition. If two or more actions are required, place each on a separate line.

In the quick-start example, we entered the state label by typing it in the state box directly. An alternative is to open the state dialog box by right-clicking the state and choosing **Properties** from the drop-down menu. Figure 14.12 shows the state dialog box for the OnZero state in the quick-start example. For more information on the state dialog box, refer to the online help system.

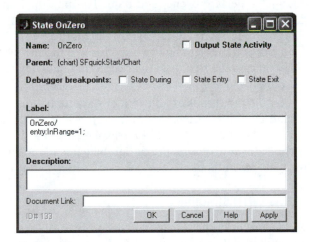

FIGURE 14.12: State dialog box

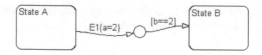

FIGURE 14.13: Condition action

14.7.2   Transition Labels

Transition labels may have up to four parts. The general format of a transition label is

$$event[condition]\{condition\ action\}/transition\ action$$

where `event` is the name of the event that triggers the transition. If no event is specified, the transition occurs immediately, assuming `condition` is true. If the transition occurs (the complete transition in the case of a compound transition), transition action is taken.

`condition action` is an action that takes place if `event` occurs and `condition` is true, whether or not the transition takes place. `condition action` is useful in the case of compound transitions, where it is necessary for an action to take place as a result of an event, even when no transition occurs. For example, consider the Stateflow chart in Figure 14.13. Here, if event E1 occurs while the system is in State A, the value of a is set to 2. But, if the value of b is not 2, the transition from State A to State B will not take place.

14.8   Stateflow EXPLORER

In the quick-start example, we defined events and variables using the **Add** menu on the Stateflow window menu bar. The Stateflow Explorer provides a means with which to view all events and data in a single window, to edit the properties of events and data, and to declare events and data. To open the Stateflow Explorer, choose **Tools:Explore** from the Stateflow window menu bar. For example, the Explorer window for the quick-start model is shown in Figure 14.14. The window contains two panes. The pane on the left allows you to browse the hierarchy of all open Stateflow charts. The pane on the right lists the events and data defined for the object selected in the left pane.

To edit the properties of an event or data item, double-click the item or select the item and choose **Edit:Properties** from the Explorer menu bar. The properties dialog box for the selected item will be displayed. To define a new event or variable, you can choose **Add** from the menu bar, which is identical to the add menu on the Stateflow window menu bar.

14.9   INTERFACING WITH Simulink

In the quick-start example, we defined events in Stateflow to be input from Simulink and also defined data to be output to Simulink. Stateflow provides many other options for interfacing between Simulink and Stateflow charts.

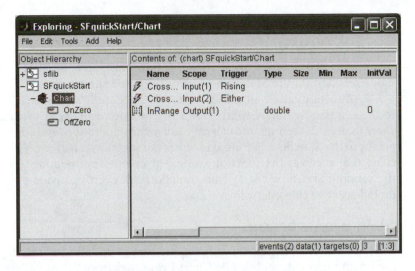

FIGURE 14.14: Stateflow Explorer window

14.9.1   Inputs

There are two basic types of inputs from Simulink to Stateflow charts. The first type of input is an event. To define an input event, choose **Add:Event:Input from Simulink** from the Stateflow or Explorer window menu bar. An event input from Simulink can be either a zero-crossing event or a function call. In the quick-start example, we created zero-crossing events using Hit crossing blocks. If the event is a zero crossing, it is not necessary to use a Zero-crossing block, because the Stateflow chart will by default exhibit automatic zero-crossing detection. However, if the event is triggered by some other value of a signal, the Zero-crossing block is necessary.

A function call event is produced in one Stateflow chart and transmitted to another Stateflow chart in the same Simulink model via a Simulink signal line.

All events are input to a Stateflow chart via a single input port. If there is more than one event, the events must be combined into a vector signal. The order of elements of the vector input signal must correspond to the index order of the events defined in Stateflow. Note that it is possible to edit the input order in Stateflow using the Stateflow Explorer by editing the index in the event dialog box.

Stateflow can also receive data inputs from Simulink. Stateflow produces an input port for each data input. The data may be vector or scalar. To define an input, choose **Add:Data:Input from Simulink** from the Stateflow window menu bar.

14.9.2   Outputs

Stateflow charts can produce two types of outputs: data and events. As we saw in the quick-start example, data outputs are defined by choosing **Add:DataOutput to Simulink** from the Stateflow or Explorer window menu bar. Data outputs can be scalar or vector. Each data output is associated with an output port, and the output port is labeled with the variable name.

A Stateflow chart can also produce output events. To define an output event, choose **Add:Event:Output to Simulink** from the Stateflow or Explorer window menu bar. Output events may be configured to trigger on either edge or as function call events.

Trigger events may be used to trigger other Stateflow charts or triggered subsystems. The output event trigger mode must be compatible with the trigger mode selected for the Stateflow chart or triggered subsystem.

A function call event causes the receiving Stateflow chart to act as a subroutine to the sending Stateflow chart. When the function call event occurs, the Stateflow chart that sends the function call event suspends operation, and control immediately passes to the Stateflow chart that receives the function call event. When the Stateflow chart that receives the function call event completes updating, control returns to the sending Stateflow chart. Thus, function call events provide a natural means to model reactive subsystems.

14.10   CHART UPDATE OPTIONS

There are three options to specify the method of updating a Stateflow chart. To set the update method, choose **File:Chart Properties** from the Stateflow window menu bar, opening the Chart Properties dialog box shown in Figure 14.15. The Chart Properties dialog box field **Update method** provides three options: `Triggered or Inherited`, `Sampled`, and `Continuous`. We'll discuss each option next.

The default update method is `Triggered or Inherited`. This option causes the Stateflow chart to update based on the chart inputs. Recall that the two types of inputs are events, which enter the chart through a trigger port, and data, which enter the chart through data ports. Events cause the chart to update each time they occur. Input data cause the chart to update at the sample rate of the signal. Therefore, if

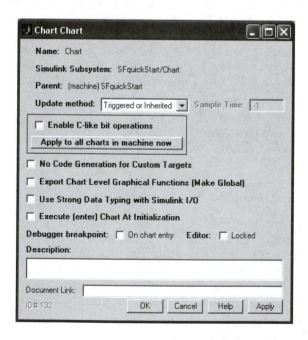

FIGURE 14.15: Chart Properties dialog box

the data signal is sampled, the chart will update each sample time. If the input is continuous, the chart will update each integration step.

If the update method is set to `Sampled`, the chart will update at intervals specified in field **Sample time**. This update method may be useful if it is necessary to update a chart at a rate different from the sample time of the input signal.

If the update method is set to `Continuous`, the chart will update at each integration step, including intermediate steps. Usually, `Triggered` or `Inherited` and `Continuous` produce the same results. The difference between the two methods is that if `Continuous` is selected, Simulink maintains an extra copy of the chart data. This allows Simulink to locate zero crossings precisely and therefore to perform associated chart updates more precisely. For example, if there is a `during` action in a state box that produces output that is a function of a current chart input, `Continuous` update should be used.

EXAMPLE 14.1   Yo-Yo Control

A tethered satellite system consists of a small satellite attached by a long tether to an orbiting platform such as the Space Shuttle. The tether can be as long as 30,000 m. One of the control problems faced by tethered satellite designers is elimination of libration (swinging) of the tethered satellite without expending the small amount of propellant carried aboard the satellite. An interesting approach to this problem is called yo-yo control [2]. We can illustrate the technique via the simpler problem of stabilizing a long pendulum.

Suppose a pendulum consists of a 100 m flexible tether supporting a 1 kg bob, as illustrated in Figure 14.16. The top of the pendulum is attached to a reel that can rapidly extend or retract the pendulum. The outer limit of the tether is 105 m, and the inner limit is 95 m. When the reel is extending or retracting the tether, the rate of change of tether length is constant at 20 m/sec. The equation of motion of the pendulum is

$$m_p l^2 \ddot{\theta} + 2m_p l \, \dot{l} \dot{\theta} + m_p g \, l \sin \theta = 0$$

where m_p represents the mass of the satellite, l represents the instantaneous tether length, and θ represents the pendulum deflection angle.

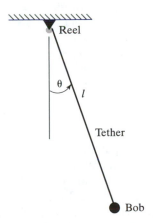

FIGURE 14.16: Yo-yo system

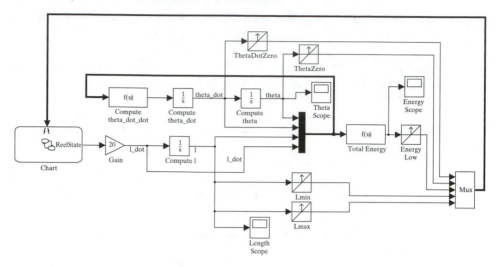

FIGURE 14.17: Yo-yo control model

Examining the equation of motion, it is evident that when the angular velocity $\dot{\theta}$ is zero, l does not affect the angular acceleration. When θ is zero, $\ddot{\theta}$ reaches its largest magnitude, and l has its maximum effect on angular acceleration. Yo-yo control is based on the following strategy: When θ reaches zero, rapidly extend the tether to its maximum length, thus decreasing angular acceleration. Then, when $\dot{\theta}$ reaches zero, retract the tether to its inner limit. Because l does not affect the angular acceleration at this point, this retraction does not increase angular acceleration. Repeat this process until the libration is reduced to an acceptable value.

The yo-yo control algorithm could be implemented using standard Simulink blocks, but it would be cumbersome. However, this control law defines a simple finite state machine and can therefore be readily modeled using Stateflow. The Simulink model in Figure 14.17 implements the equation of motion and produces hit crossing events when $\theta = 0$ and when $\dot{\theta} = 0$. The model produces additional hit crossing events when the tether reaches its inner and outer limits. The model also produces a hit crossing event when the total energy (sum or kinetic and potential energy) crosses a threshold, set to 0.001. The control logic is modeled in the Stateflow chart shown in Figure 14.18. A typical trajectory is shown in Figure 14.19. The upper plot shows the history of θ, and the lower plot shows the history of tether length.

An interesting exercise would be to enhance the model such that it can detect when the total energy exceeds some threshold and can restart the yo-yo process.

14.11   FLOWCHARTS

Stateflow provides the capability to represent algorithmic functions via flowcharts. The flowcharts are implemented as compound transitions. This capability provides the means to model standard flowchart features, including decision making (if..then..else) and looping. In this section, we will show how to implement a flowchart Stateflow chart and each of the flowchart elements.

A flowchart can be incorporated in any Stateflow chart. However, it may be convenient to implement a flowchart as a stand-alone subroutine. The basic structure

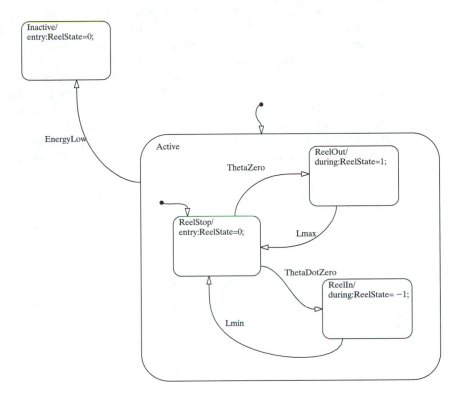

FIGURE 14.18: Yo-yo control Stateflow chart

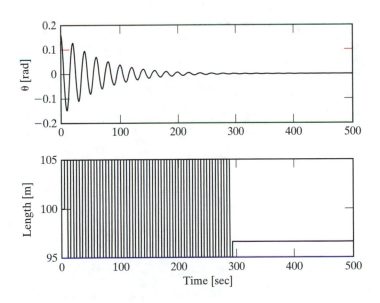

FIGURE 14.19: Yo-yo control model trajectories

of a Stateflow flowchart block is shown in Figure 14.20 and Figure 14.21. The first time the Stateflow chart is triggered, the default transition to the starting junction takes place. Next, the flowchart is executed by the evaluation of the compound transition from the starting junction to the end state. If control reaches the end state, each subsequent activation of the flowchart begins with the inner transition to the starting junction and completes at the end state. The two main flowchart components are decision making (Figure 14.20) and looping (Figure 14.21).

A decision-making segment of a flowchart consists of at least one conditional transition and one unconditional transition (the else clause). The conditional branch transition label contains a condition (in square brackets []) and optionally, a

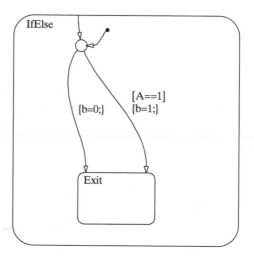

FIGURE 14.20: Decision-making flowchart

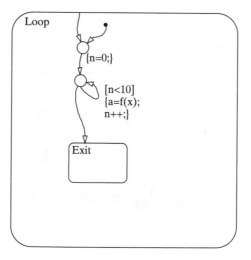

FIGURE 14.21: Looping flowchart

condition action in curly braces ({ }). Figure 14.20 illustrates a simple flowchart with one decision step. In this flowchart, each time the Stateflow chart is activated, the value of A is tested. If A is 1, b is set to 1. Otherwise, b is set to 0.

Notice that the assignment statements in condition actions are terminated with semicolons. If a semicolon is not present, the results of the assignment will be displayed in the MATLAB workspace each time the flowchart executes. The condition action can contain multiple assignment statements, separated by semicolons. Also, notice that the condition and condition action can be separated by a carriage return, and multiple assignments in a condition action can also be separated by carriage returns (but the semicolons are still required).

Figure 14.21 illustrates the structure of a loop. The loop contains two transitions. The first is the self loop from the connective junction back to itself. This transition is repeated as long as the condition is true, and each iteration, the condition action takes place. When the condition is no longer true, the transition without the condition takes place. There can be additional condition actions on this transition.

EXAMPLE 14.2   Newton iteration flowchart

Suppose we wish to implement a Simulink block that computes a root of

$$a x^3 + 2x + 1 = 0$$

Coefficient a is the block input, and real root x is the block output. A Stateflow flowchart that performs this task using Newton–Raphson iteration (see Section 13.3) is shown in Figure 14.22. The block is incorporated in the Simulink model shown in Figure 14.23. In this

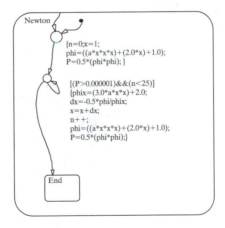

FIGURE 14.22: Newton iteration Stateflow flowchart

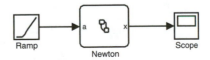

FIGURE 14.23: Simulink model incorporating Newton iteration flowchart

flowchart, a is defined to be an input, x is defined to be an output, and all other variables are defined to be local. Therefore, the Stateflow block has one scalar input port and one scalar output port. The block is configured to update continuously, so x is computed each simulation step.

14.12 GRAPHICAL FUNCTIONS

Graphical functions provide the capability to create state boxes in Stateflow that serve the same purpose as MATLAB functions or S-functions. A graphical function can be thought of as an internal subroutine or function in a Stateflow chart. Graphical functions accept input arguments and return values. Graphical functions can be used in any Stateflow chart, but they are especially useful in flowcharts.

To illustrate the structure of graphical functions, consider the chart shown in Figure 14.24. This chart exhibits the same behavior as the loop example in Figure 14.21. Notice that the graphical function boxes have square corners, and the box labels are function prototypes that specify the function names, input arguments, and return variables.

To create a graphical function, drag a state box from the drawing palette. Right-click the state box, and choose **Type:Function**. The state box will be transformed into a graphical function box with square corners. Also, the function prototype will be inserted as the graphical function box label. Edit the label to add the function name and input and output arguments. The graphical function should contain, as a minimum, a default transition terminated at a junction.

The input and output arguments in a graphical function are formal arguments. Therefore, they do not have to be the same as the variable names used to invoke the function. Graphical functions do not accept array arguments.

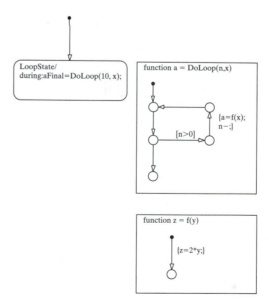

FIGURE 14.24: Loop example using graphical functions

14.13 PRINTING Stateflow CHARTS

There are two ways to print a Stateflow chart. The easier is to use the **File:Print** menu choice on the Stateflow window menu bar. The second method is to use the MATLAB command `sfprint`.

The Stateflow print dialog box has options for printing the current view of the Stateflow chart or printing the entire chart. The chart can be sent directly to the printer or saved in a file.

The `sfprint` command has two syntax variations. The first is

```
sfprint
```

This version of the command is equivalent to choosing the **File:Print** from the Stateflow window menu bar. The second version is

```
sfprint(chart name, format, [output option],
[PrintEntireChart])
```

where *chart name* is a MATLAB string containing the name of the Stateflow chart (the name of the chart block in the Simulink model), and *format* is a MATLAB string containing an output format from Table 14.3. *output option* is an optional parameter from Table 14.4. The default behavior is for the image of the current Stateflow chart to be stored in file *chart name.format*. *PrintEntireChart* is an optional parameter that may contain the value 0 or 1. If *PrintEntireChart* is 0, the current view of the chart is printed. If *PrintEntireChart* is 1 (the default), the entire chart is printed.

14.14 Stateflow DEBUGGER

The Stateflow debugger provides a convenient means to monitor the execution of a Stateflow chart. To start the debugger, select **Tools:Debug** from the Stateflow window menu bar. The Stateflow debugger is shown in Figure 14.25. Operation of the Stateflow debugger is similar to operation of the Simulink debugger. You can set breakpoints for chart entry, event broadcast, or entry into a selected state. Once a break occurs, you can single step through the Stateflow chart operation or continue to the next breakpoint. The display area at the bottom of the Stateflow debugger

TABLE 14.3: `sfprint` Format Options

Format	Meaning
'eps'	Encapsulated PostScript
'epsc'	Color Encapsulated PostScript
'ps'	PostScript
'psc'	Color PostScript
'tif'	Tagged image format
'jpg'	jpeg format
'png'	PNG format
'meta'	Save to the clipboard (PC or Mac only) in PC metafile format
'bitmap'	Save to the clipboard (PC or Mac only) in PC bitmap format

TABLE 14.4: `sfprint` Output Options

Output Option	Result
filename	Save image in specified file
'promptForFile'	Filename dialog box will be displayed
'printer'	Send output to printer (only available with PostScript formats)
'file'	Save in file *ChartName.format*, where *ChartName* is the full name of the chart, and *format* is the filename extension associated with the specified format
'clipboard'	Save output to the clipboard

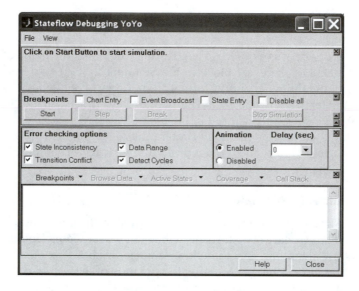

FIGURE 14.25: Stateflow debugger window

window provides useful debugging information. For details on the operation of the Stateflow debugger, see the online documentation.

14.15 SUMMARY

In this chapter, we discussed finite state machines and the Statechart notation. We also provided a brief introduction to Stateflow. We presented a quick-start tutorial that allows you to go through the complete process of defining and using a Stateflow chart. We also described the basics of defining events and data and interfacing Stateflow charts with Simulink models. Finally, we discussed printing Stateflow charts.

This introduction was necessarily brief. Stateflow is a powerful programming environment, and it has many features that we did not discuss. Once you master the basics described here, you may find it beneficial to review Ref. [7] and the online documentation. You may also find it helpful to study the examples included with Stateflow.

REFERENCES

The best way to learn more about applications of Stateflow is to review the technical literature on Statecharts. Although there are some syntactical differences between Stateflow and Statecharts, the examples in the Statechart literature are helpful. The papers listed below and their references should provide a good start.

[1] Caryl, Matthew, Statecharts, http://www.catachan.demon.co.uk/Alife/state-charts.html. This is an online Statecharts tutorial oriented toward artificial life programs.

[2] Dabney, James B., Hasdorff, Lawrence, Harman, Thomas L., and Watson, James T., "Effects of Tether Stretch Dynamics on a Tethered Satellite Yo-yo Control Law," Joint Applications in Instrumentation Processes and Computer Control, Houston, Texas, 1993.

[3] Harel, David, "On Visual Formalisms," *Communications of the ACM* 31 (5, May 1988): 514–530. This paper presents an overview of hierarchical visual formalisms and a clear description of Statecharts.

[4] Harel, David, and Gery, Eran, "Executable Object Modeling with Statecharts," Proceedings of the 18th International Conference on Software Engineering (IEEE Press, March 1996): 246–257. This paper presents an application of Statecharts for modeling object-oriented systems. It includes an example model of a railcar switching system.

[5] Harel, David, and Naamad, Amnon, "The Statemate Semantics of Statecharts," *ACM Transactions on Software Engineering and Methodology* 5 (4, October 1996): 293–333. This paper presents detailed semantics of Statecharts.

[6] *Stateflow User's Guide* (Natick, Mass.: The MathWorks, Inc. 2002). The Stateflow user's guide provides detailed information on all aspects of Stateflow.

[7] Paulo, Fabiano Borges, Masiero, Paulo Caesar, de Oliveira, Maria Cristina Ferreira, "Hypercharts: Extended Statecharts to Support Hypermedia Specification," *IEEE Transactions on Software Engineering* 25 (1, January/February 1999): 33–49. This paper presents an interesting application of Statecharts to model hypermedia.

15

Introduction to Real-Time Workshop

In this chapter, we will discuss Real-Time Workshop. We'll begin with an overview of Real-Time Workshop, including a description of targets and the build process. Then, we will illustrate the use of Real-Time Workshop with examples of two of the many possible operational scenarios.

15.1 INTRODUCTION

Real-Time Workshop is an extension to Simulink that translates Simulink models into executable code. Real-Time Workshop can be configured to produce executable code for the computer on which Simulink is running, another PC, a digital signal processing card, or even an embedded controller. Real-Time Workshop is useful for a variety of applications. It can be used to produce a model that executes much faster than the original Simulink model, facilitating optimization and Monte Carlo analysis. It can also be used to build executable programs for distribution or for use in real-time controllers.

This chapter presents an introduction to Real-Time Workshop and demonstrates two of the many feasible development scenarios using Real-Time Workshop. We will start with an overview of targets and the build process using the generic real-time target. Next, we will demonstrate the use of Real-Time Workshop for fast simulation using the rapid simulation target supplied with Real-Time Workshop. Finally, we will demonstrate the development of an external target using xPC.

15.1.1 Development Scenarios

Real-Time Workshop generally involves two computers, the *host* and the *target*. The *host* is the computer on which MATLAB and Simulink are executed. The *target* computer is the computer on which the Real-Time Workshop executable will be loaded. The target can be physically the same computer as the host, such as when the rapid simulation capability is used. The target can also be another computer connected to the host or it can be a real-time controller such as an embedded microprocessor.

Real-Time Workshop is useful for a variety of operational scenarios. One important application we will demonstrate is rapid simulation. Because Real-Time Workshop builds executable programs, they can execute much faster than Simulink models running on the host. The rapid simulation capability can be exploited to perform Monte Carlo analysis, optimization, or sensitivity analysis. The rapid simulation scenario consists generally of the following steps:

1. Develop and test a Simulink model of the process. It is convenient to specify the parameters to be varied as MATLAB variables.

2. Build a Rapid Simulation target, and test it in comparison to the original Simulink model.

3. Develop a function M-file that sets parameters based on one or more inputs, executes the rapid simulation target, and returns results.

4. Develop a driver M-file that performs the desired analysis.

A second operational scenario is rapid application development (RAD), as discussed in Section 1.3.2. For example, suppose we wish to develop the software for a real-time embedded controller. Using Real-Time Workshop, we could proceed as follows:

1. Build a Simulink subsystem that models the system to be controlled.

2. Build a Simulink subsystem that implements the control algorithm.

3. Test the controller in Simulink, and tune parameters as necessary.

4. Build the target using Real-Time Workshop, and download it to the target computer.

5. Test and validate the target using the actual system to be controlled, and deploy the product.

15.1.2   Build Process

Real-Time Workshop adds several submenus to the Simulink **Simulation:Parameters** dialog box and the Simulink **Tools** menu. Targets are built using the **Real-Time Workshop** page of the **Simulation:Parameters** dialog box. This page also allows you to select the target and set target options. For certain real-time targets, such as xPC, the build process also automatically downloads the target application to the target computer and places the target application in a wait mode.

The automatic build process controlled by Real-Time Workshop consists of the following steps:

1. Translate Simulink model into intermediate and rtw format file.

2. Convert the rtw file into C code for the target C compiler.

3. Generate a custom make file.

4. Execute the make file to compile and link the target application.

5. If appropriate, download the target application and start it on the target computer.

We will illustrate the entire process with three examples using the Generic Real-Time target, Rapid Simulation target, and xPC.

15.2 GENERIC REAL-TIME TARGET

Generic Real-Time (GRT) target is supplied with Real-Time Workshop. GRT permits you to build executable models that you can run on any computer using the same processor type and operating system. GRT is useful as an intermediate step between executing a model entirely from within Simulink and executing it on an embedded processor. GRT also supports External Mode, which allows you to control the executing program from Simulink via TCP/IP and display model output using Simulink sinks such as Scope blocks. External mode targets can be executed on the host computer or on any other computer accessible via TCP/IP.

To illustrate building and using a GRT target on a Windows platform, let's start with the simple spring-mass-dashpot model shown in Figure 15.1. After building the model, we can test it as a standard Simulink program, tune parameters, and display results.

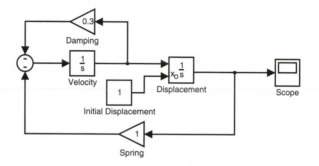

FIGURE 15.1: Simulink model of spring-mass-dashpot system

15.2.1 Stand-Alone GRT Target

Next, we generate a stand-alone GRT target using Real-Time Workshop. When Real-Time Workshop builds an executable program from the model, the executable program saves all data that the model sends to the MATLAB workspace in a .mat file. Recall that there are a number of ways to save data to the MATLAB workspace. **Simulation:Parameters** page **Workspace I/O** allows you to save time, states, output, and the final state. You can also use To Workspace blocks or use the **Data history** page of the Scope block **Parameters** dialog box as illustrated here.

Configure the Scope block **Data history** page as shown.

Next, we will use the **Simulation:Parameters** dialog box to configure the model such that it's compatible with Real-Time Workshop and then to build the GRT target.

Select the **Solver** page, and choose a fixed-step solver and appropriate stepsize. (GRT does not support variable-step size solvers.)

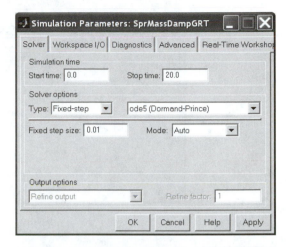

Open the Real-Time Workshop page and click the **Browse** button to open the System Target File Browser shown in Figure 15.2. Choose target grt.tlc. The configuration fields should automatically be filled as shown here.

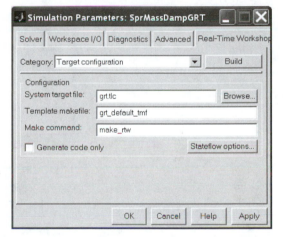

Click the **Build** button. The MATLAB command window will be displayed, and a sequence of status messages will be displayed as the GRT target is built. The target will be saved as an executable file with the same name as the Simulink model and extension .exe. We saved the model as SprMassDampGRT.mdl, so the executable program is saved as SprMassDampGRT.exe.

15.2.2   Running the Stand-Alone GRT Model

To run the model, open a DOS window and enter the program name at the DOS prompt. The model runs to completion and displays a message as shown in Figure 15.3.

Once the model completes execution, all model output is saved in a .mat file with the same name as the model. The .mat file can be loaded into MATLAB and postprocessed or plotted. For example, in this case, we save the Scope data

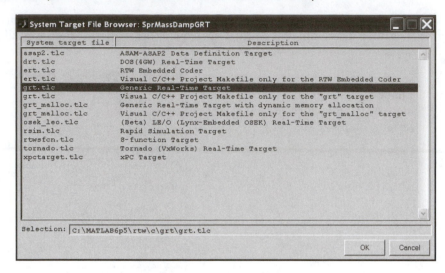

FIGURE 15.2: Selecting a target

FIGURE 15.3: Executing a target

as a structure with time in variable `Position`. Load the `.mat` file with MATLAB command

```
load SprMassDampGRT
```

The MATLAB workspace now contains an additional structure `rt_Position`. Notice that the GRT executable adds prefix `rt_` to the output variable name. Listing 15.1 loads the `.mat` file and produces a plot of mass position.

Listing 15.1: Accessing GRT output

```
%Load GRT output and plot data
load SprMassDampGRT
 t =  rt_Position.time ;
 x = rt_Position.signals(1).values ;
 plot(t,x) ;
 xlabel('Time [sec]');
 ylabel('Position') ;
```

15.2.3 Using External Mode

External mode allows you to control execution of the GRT model from within Simulink. The GRT model can be run on the host computer or on another computer connected to the host via TCP/IP. To use external mode, the GRT target must be rebuilt.

Open the **Simulation:Parameters Real-Time Workshop** page and category GRT code generation options. Select check box **External mode** and rebuild the GRT executable.

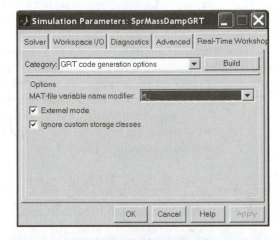

From the model window menu bar, choose **Tools:External mode control panel**. Click button **Signal & triggering** to open the External Signal and Triggering window.

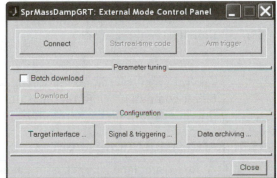

Verify that the output signal is selected and set field **Duration** to a value greater or equal to the number of time points in the output. Close this window.

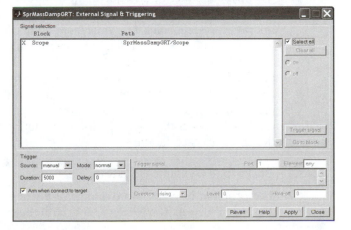

Load the executable program in a DOS window by entering the program name at the DOS prompt with command line option -w. For the example program here, the command is

```
SprMassDampGRT -w
```

The program will load and immediately enter a waiting condition.

Next, choose **Simulation:External** from the model window menu bar. Then, press the **Connect** button on the External Mode Control Panel (**Tools:External mode control panel**. Choose **Simulation:Start real-time code**. The model will execute, and the output will be displayed on the Scope and should be identical to the output produced earlier. To run the model in Simulink instead of external mode, choose **Simulation:Normal**.

To run the target in external mode on a computer connected to the host via TCP/IP (typically a local area network connection), the process is nearly the same. Copy the executable program to the remote target computer and start it in a DOS window using the -w option. In the External Mode Control Panel, click the **Target interface** button.

Enter in field **MEX-file arguments** the network name of the target computer or the IP address as a MATLAB string (in single quotes). Then start the program on the target using **Simulation:Start real-time code**.

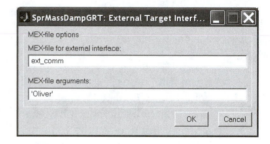

The program will run on the target computer and save the output .mat file on the target computer. While the program is running, all displays and controls (such as the slider gain) will be active in the Simulink model window.

15.3   RAPID SIMULATION TARGET

A rapid simulation target is a stand-alone executable program that operates on the host computer or an equivalent computer (same processor type and operating system). It is possible to execute the simulation target from within MATLAB or an operating system batch file, thus facilitating Monte Carlo analysis or optimization. A rapid simulation target does not operate in real time.

We will illustrate using the rapid simulation target with a minimization example. Consider the system shown in Figure 15.4. The system consists of a cannon that fires basketballs at a specified angle and a moving cart with a basket target. The problem we wish to solve is to determine a proper angle for the cannon barrel such that the ball goes through the basket, assuming the cannon is fired at the same time the cart starts accelerating.

15.3.1   Basketball Dynamical System

The problem involves two dynamical subsystems: the cart and the basketball. The 2 kg cart is driven by a force of 20 N until the cart reaches a speed of 5 m/s. The cart

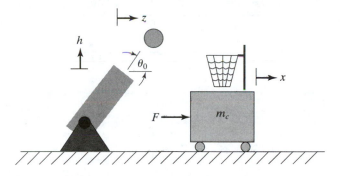

FIGURE 15.4: Basketball cannon system

then continues moving at 5 m/s. The cart equation of motion is

$$\ddot{x} = F/m_c$$

where

$$F = 20 \text{ N}, \qquad\qquad \dot{x} < 5 \text{ m/sec}$$

$$= 0, \qquad\qquad \dot{x} \geq 5 \text{ m/sec}$$

and m_c is the cart mass.

The basketball is fired at a velocity of 15 m/sec and is affected by gravity and aerodynamic drag. The initial height of the basketball is 0.5 m above the basket. The equation of motion of the basketball in the horizontal direction (coordinate z here) is

$$\ddot{z} = -F_d \cos \theta / m_b$$

and in the vertical direction (coordinate h)

$$\ddot{h} = g - F_d \sin \theta / m_b$$

where F_d is the aerodynamic drag

$$F_d = \frac{1}{2} C_d A_b \rho v^2$$

and v is the basketball speed, C_d is the drag coefficient taken to be 1.0, A_b is cross-sectional area of the 0.15 m radius ball, and ρ is the air density (1.224 kg/m^2). θ is the instantaneous flight path angle of the basketball,

$$\theta = \tan^{-1}(\dot{h}/\dot{z})$$

15.3.2   Basketball Simulink Model

A Simulink model that implements this dynamical system is shown in Figure 15.5. The various parameters are set in the M-file in Listing 15.2. The simulation is configured to stop when the ball passes the basket in the vertical plane ($h = 0$).

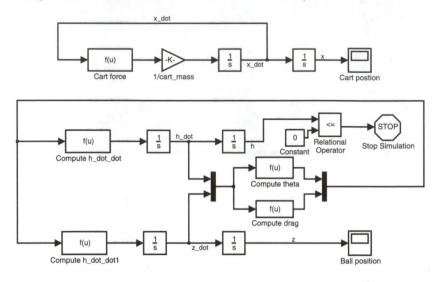

FIGURE 15.5: Basketball Simulink model

Listing 15.2: Parameter initialization script

```
%Initialization for basketball game
F0 = 20.0 ; % N
cart_mass = 2 ; %kg
x_dot_max = 5 ; % m/sec
ro_air = 1.224 ; % kg/m^3
Cd = 1 ;
r_ball = 0.15 ; % m
A_ball = pi*r_ball^2 ;
AeroFac = Cd*A_ball*ro_air/2 ;
ball_mass = 0.4 ; % kg
g = -9.8 ; % m/sec^2
theta_0 = pi/2.5 ; % rad
v0 = 15 ; % m/sec
```

The Scope blocks are configured to send the output to the MATLAB workspace. The cart position is stored in matrix `CartPos`, and the basketball horizontal position is stored in matrix `BallPos`. All parameters, including the shot angle θ_0 are set by M-file `BasketBallInit.m`. Once the parameters are set, the model can be run like any other Simulink model.

When the program stops, matrices `CartPos` and `BallPos` will be available in the MATLAB workspace. Each matrix will contain two columns. The first is simulation time and the second the position.

15.3.3   Building the Rapid Simulation Target

The target is built using the **Simulation:Parameters** dialog box. The rapid simulation target requires a fixed step size solver, so we will use ODE5 with a step size of 0.01 sec.

First, choose the **Simula-tion:Parameters Solver** page. Choose `Fixed-step`, `ODE5`, and set `Fixed step size` to 0.01.

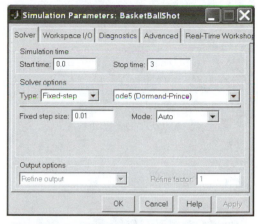

Next, choose the **Real-Time Work-shop** page and set **Category** to `Target configuration`. Using the **Browse** button, select **System target file** `rsim.tlc`. The remaining fields should automatically appear as shown.

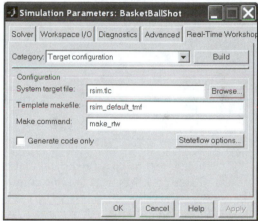

If you have not run M-file `BasketBallInit.m`, run it from the MATLAB prompt. Next, press the **Build** button. The MATLAB command window will pop up and display a series of messages as the rapid simulation executable program is built. When the successful completion message is displayed, there will be a new executable program (for example, `BasketBallShot.exe` in Windows) in the current MATLAB directory.

You can run the executable program from a command window or from within MATLAB. To run it from within MATLAB, enter the command

```
!BasketBallShot
```

at the MATLAB prompt. The program will run and store the output trajectory in file `BasketBallShot.mat`.

To plot the trajectory, first load `BasketBallShot.mat`, which contains the two trajectory matrices `CartPos` and `BallPos`. However, the matrices will have prefix `rt_`, so they will be named `rt_CartPos` and `rt_BallPos`.

15.3.4   Rapid Simulation Options

The rapid simulation target executable is built to respond to several optional command line arguments. These options allow you to specify input and output

TABLE 15.1: Rapid Simulation Arguments

Argument	Meaning
-f *OldInput*.mat=*NewInput*.mat	Substitute input signal matrix file *NewInput*.mat for *OldInput*.mat, where *OldInput*.mat is the file specified in a From File source block.
-o *outputfile*.mat	Send the trajectory output to file *outputfile*.mat rather than the default *model*.mat.
-p *ParmFile*.mat	Read model parameters from file *ParmFile*.mat, overriding the values stored with the model.
-s *stoptime*	Override the simulation stop time.
-t *OldOutput*.mat=*NewOutput*.mat	For a To File block configured to store data in *OldOutput*.mat, instead store the data in *NewOutput*.mat.
-v	Verbose mode. Causes the rapid simulation target to print more status messages as the simulation executes.
-h	Display a help screen listing the available command line arguments.

filenames, to change the simulation stop time, and to change model parameters. Table 15.1 lists the available command line arguments.

Signal Files. Two of the command line arguments allow you to change the names of signal files. Option -f selects a different input file for a From File block, and -t selects a different output file for a To File block. The signal file formats are as documented for the From File and To File blocks.

Parameter Files. Recall that when you execute a Simulink model, any block parameters that contain variables defined in the MATLAB workspace are evaluated. This capability allows you to tune parameters, perform optimizations, or use Monte Carlo techniques. A rapid simulation target provides a similar mechanism via the -p command line option. All rapid simulation models parameters are stored in a structure that you can change using MATLAB command rsimgetrtp. The process is as follows:

1. Assign values to all parameters you wish to change. The parameters must be assigned in the base workspace. If you are using a function M-file, use command assignin.

2. Create a parameter structure using rsimgetrtp. For example, to create a parameter structure named Newrtp for model MyModel.exe, enter the command

```
Newrtp = rsimgetrtp('MyModel') ;
```

3. Save the parameter structure as a .mat file using MATLAB command save.

4. Execute the rapid simulation model using the -p option.

The use of the -p option is illustrated in the next section.

15.3.5 Minimizing Miss Distance

For our basketball cannon, we wish to adjust the firing angle θ_0 such that the ball goes through the basket. Therefore, we wish to minimize the magnitude of the difference in x and z at the instant the simulation stops. We can solve this problem using MATLAB function fminsearch. We will need two M-files. The first is an objective function that takes as its argument a trial value of θ_0 and returns the square of the miss distance. The second is a script M-file that uses fminsearch to find the shot angle and prints the result.

A suitable objective function M-file is shown in Listing 15.3. Note that all model parameters are set in the base workspace, and then rsimgetrtp is used to construct a parameter structure. Next, the model is executed using the -p option. The model saves the Scope block output trajectories in .mat file BasketBallShot.mat, which is loaded into the MATLAB workspace using MATLAB command load. The final values of x and z are recovered from arrays rt_CartPos and rt_BallPos and used to compute the return value.

Listing 15.3: Objective function M-file

```
function P = MissDistance(theta_0)
%Compute miss distance squared for specified shot angle.
%This serves as the objective function for minimization using
%fminsearch.
%Parameters
F0 = 20.0 ; % N cart force
cart_mass = 2 ; %kg
x_dot_max = 5 ; % m/sec  final cart speed
ro_air = 1.224 ; % kg/m^3
Cd = 1 ;
r_ball = 0.15 ; % m
A_ball = pi*r_ball^2 ;
AeroFac = Cd*A_ball*ro_air/2 ;
ball_mass = 0.4 ; % kg
g = -9.8 ; % m/sec^2
v0 = 15 ; % m/sec ball initial speed
%rsimgetrtp requires parameters to be set in the base workspace.
assignin('base', 'F0',F0) ;
assignin('base', 'cart_mass',cart_mass) ;
assignin('base', 'x_dot_max',x_dot_max) ;
assignin('base', 'AeroFac',AeroFac) ;
assignin('base', 'ball_mass',ball_mass) ;
assignin('base', 'g',g) ;
assignin('base', 'v0',v0) ;
assignin('base', 'theta_0',theta_0) ;
Newrtp = rsimgetrtp('BasketBallShot') ;
save ShotParams Newrtp ;
!BasketBallShot -p ShotParams.mat
load BasketBallShot ;
np = max(size(rt_CartPos)) ;
xf = rt_CartPos(np,2) ;
zf = rt_BallPos(np,2) ;
P = (xf - zf)^2 ;
```

A script M-file that finds the shot angle using `fminsearch` is shown in Listing 15.4. This script computes a shot angle of 55.031250 deg.

Listing 15.4: Finding the shot angle

```
%Script to find the shot angle required to make a basket,
%shooting the basketball cannon at a moving basket.
InitialGuess = pi/3 ;
thetaShoot = fminsearch('MissDistance',InitialGuess)*180/pi ;
fprintf('\nShoot at %f deg', thetaShoot) ;
```

15.4   xPC TARGET

A typical Real-Time Workshop target is a digital signal processing board or an embedded controller. xPC is a target available from The MathWorks that runs on any PC. In particular, you can use an older PC, even one based on an 80486 CPU, thus providing an inexpensive real-time system. If necessary, inexpensive input/output boards or A/D and D/A interface boards can be added to the target PC. Additionally, xPC requires either the Watcom C compiler or Microsoft Visual C.

The target PC is connected to the host via either a serial port connection using a cable supplied with xPC or via TCP/IP using any of several standard Ethernet adapters. xPC is also supplied with a PCI bus Ethernet card for the target PC that is compatible with xPC.

The target PC runs a special real-time operating system generated by the xPC Target Setup program. xPC Target Setup creates a boot floppy containing the target PC real-time operating system. The xPC target software is transferred to the target PC, and then the target PC executes the target software under control of the host PC. We will illustrate the complete process using the basketball cannon example.

15.4.1   Initialization

The first step in the initialization process is to create the target boot disk. Start xPC Target Setup by entering the command

```
xpsetup
```

at the MATLAB prompt. The xPC Target Setup window (Figure 15.6) will be displayed. Using this window, you can choose which C compiler you will use, configure the serial or TCP/IP connection, and build the target boot disk. For this example, we are using the serial connection, so both the host and target fields are set accordingly. Place a formatted floppy in the host computer disk drive, and click **BootDisk**.

Remove the floppy from the host computer and insert it in the target computer. Note that the target computer must be configured to boot from a floppy. Start the target computer. The target computer display will consist of a status section (Figure 15.7) and a larger window that can be used to display program output. Close the xPC Target Setup window and then enter MATLAB command `xpctest` to test the connection.

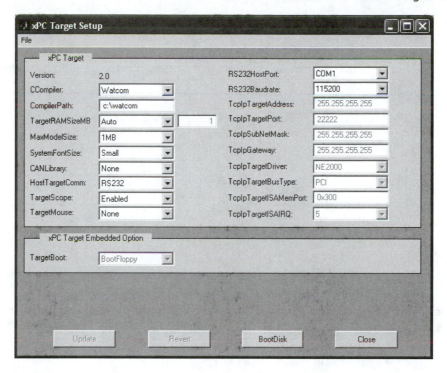

FIGURE 15.6: xPC Target Setup window

FIGURE 15.7: Booting the Target PC

15.4.2   Building a Target

Building an xPC target is similar to building a Rapid Simulation target. The **Simulation:Parameters Solver** page must be configured to use a fixed-step solver, and an appropriate step size must be chosen. The target can be configured to run as a real-time application or in free run mode, which allows execution to be as fast as possible. If real-time execution is desired, the solver step size determines the sampling rate of the real-time system and must be selected taking into consideration timing issues of the real-time system as well as computational burden on the target.

The Simulink model is modified to include logic to compute miss distance and route the miss distance to an output port. The revised model is shown in Figure 15.8.

x_dot

f(u)

Cart force

-K-

1/cart_mass

$\frac{1}{s}$ x_dot

$\frac{1}{s}$ x

Cart postion

Separation

1

Out1

f(u)

Compute h_dot_dot

$\frac{1}{s}$ h_dot

Integrator h_dot

$\frac{1}{s}$ h

0

Constant

<=

Relational
Operator

STOP

Stop Simulation

f(u)

Compute theta

f(u)

Compute drag

f(u)

Compute h_dot_dot1

$\frac{1}{s}$ z_dot

Integrator z_dot

$\frac{1}{s}$ z

Ball position

FIGURE 15.8: Simulink model modified for xPC example

Open the **Simulation:Parameters Real-Time Workshop** page and select **Category** Target configuration. Click **Browse** and choose system target file **xpc-target.tlc** from the list.

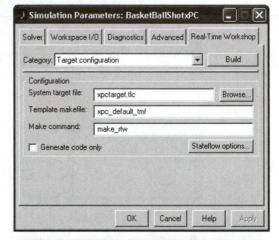

Select **Category** xPC Target code generation options (contd.) and check **External mode**.

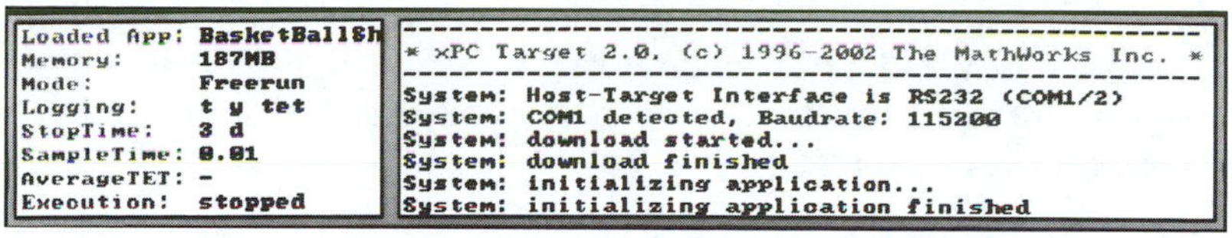

```
Loaded App: BasketBallSh   -------------------------------------------------
Memory:     187MB          * xPC Target 2.0, (c) 1996-2002 The MathWorks Inc. *
Mode:       Freerun        -------------------------------------------------
Logging:    t y tet        System: Host-Target Interface is RS232 (COM1/2)
StopTime:   3 d            System: COM1 detected, Baudrate: 115200
SampleTime: 0.01           System: download started...
AverageTET: -              System: download finished
Execution:  stopped        System: initializing application...
                           System: initializing application finished
```

FIGURE 15.9: Target loaded on Target PC

Click **Build**. The MATLAB command window will be displayed, and a sequence of messages will report the progress of building the target application and downloading it to the target. After the download is complete, the target computer screen status area will indicate that the application is loaded (Figure 15.9) and stopped.

15.4.3 Starting Target

The interface to the xPC target from within MATLAB is a data structure named tg. To start the target application, enter at the MATLAB prompt the command

```
start(tg)
```

The application will run until either it completes (reaches the specified stop time or event) or you manually stop it with MATLAB command

```
stop(tg)
```

The target PC display screen for the BasketBallShotxPC model after execution is shown in Figure 15.10. You can also execute the target using the xPC Target Control Tool (MATLAB command xpcrctool).

15.4.4 Controlling the Target

xPC provides a rich set of commands and tools with which to control and monitor the target application. For example, you can add target scopes and display signals in real time on either the host or target PC. You can replace input signals, tune parameters, and capture output signals with xPC MATLAB commands. To illustrate a few of the commands, we will find optimal basketball shot angle using xPC.

The basketball shot problem entails varying the shot angle θ_0 such that the miss distance is minimized. Therefore, we need to be able to adjust the shot angle and capture the miss distance. Unlike the Rapid Simulation target, MATLAB variables are not parameters. Instead, the block dialog box fields are the parameters for an xPC target. Therefore, instead of locating a parameter corresponding to MATLAB

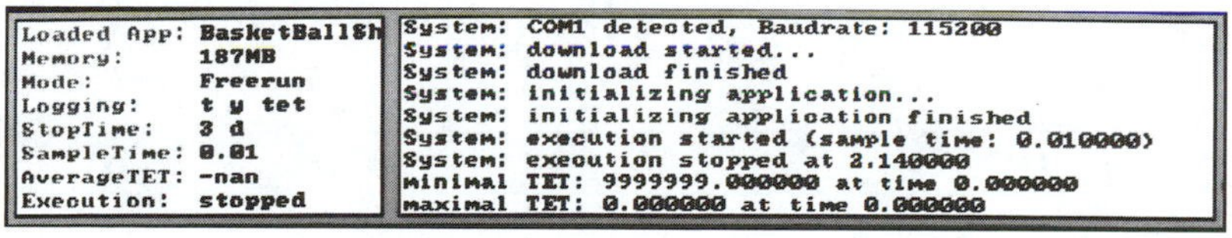

```
Loaded App: BasketBallSh   System: COM1 detected, Baudrate: 115200
Memory:     187MB          System: download started...
Mode:       Freerun        System: download finished
Logging:    t y tet        System: initializing application...
StopTime:   3 d            System: initializing application finished
SampleTime: 0.01           System: execution started (sample time: 0.010000)
AverageTET: -nan           System: execution stopped at 2.140000
Execution:  stopped        minimal TET: 9999999.000000 at time 0.000000
                           maximal TET: 0.000000 at time 0.000000
```

FIGURE 15.10: Target execution on Target PC

variable theta_0, we locate parameters corresponding to **Initial condition** fields for the h_dot and z_dot integrators. All available parameters of an xPC target are stored in the tg structure as tg.p1, tg.p2, etc. To locate the parameters, enter at the MATLAB prompt

```
set(tg,'ShowParameters','on')
```

then display the tg structure by entering its name

```
tg
```

The tg structure will then be displayed, including a table that lists the association of parameter number and parameter name. Listing 15.5 shows the commands and relevant MATLAB output (much of the output is omitted here).

Listing 15.5: xPC parameter identification

```
>> set(tg,'ShowParameters','on')
>> tg
   ShowParameters  = On
   Parameters = PROP. VALUE      TYPE   SIZE   PARAMETER NAME    BLOCK NAME
              P0     0.000000  DOUBLE Scalar InitialCondition  Integrator1
              P1     0.000000  DOUBLE Scalar InitialCondition  Integrator5
              P2     0.500000  DOUBLE Scalar InitialCondition  Integrator2
              P3     0.000000  DOUBLE Scalar Value             Constant
              P4     0.000000  DOUBLE Scalar InitialCondition  Integrator
              P5     0.500000  DOUBLE Scalar Gain              1//cart_mass
              P6    14.265800  DOUBLE Scalar InitialCondition  Integrator h_dot
              P7     4.635250  DOUBLE Scalar InitialCondition  Integrator z_dot
```

We note that parameter InitialCondition for Integrator h_dot is parameter p6 and InitialCondition for Integrator z_dot is parameter p7. We can compute the initial h_dot as

$$\dot{h}_0 = v_0 \sin \theta_0$$

and initial z_dot

$$\dot{z}_0 = v_0 \cos \theta_0$$

When we assign a value to a target parameter, the value is immediately transferred to the target. The target output is stored in tg.OutputLog and the time output in tg.TimeLog. Listing 15.6 performs a single trial given θ_0 and returns the miss distance. Listing 15.7 finds the value of θ_0 that minimizes the miss distance using MATLAB function fminsearch and the xPC target.

Listing 15.6: Computing miss distance with xPC

```
function P = MissDistancexPC(theta_0)
%BasketBallShot objective funtion for xPC
%Run BAsketBallShot once and plot output
v0 = 15 ; % m/sec ball initial speed
tg=evalin('base','tg');
tg.p6 = v0*sin(theta_0) ;
tg.p7 = v0*cos(theta_0) ;
start(tg) ;
plot(tg.TimeLog,tg.OutputLog)
yout = get(tg,'OutputLog');
np = max(size(yout)) ;
P = yout(np)^2 ;
disp(P)
pause(0.001);
```

Listing 15.7: Minimizing the miss distance

```
%Script to find the shot angle required to make a basket;
%shooting the basketball cannon at a moving basket.
InitialGuess = pi/3 ;
thetaShoot = fminsearch('MissDistancexPC',InitialGuess)*180/pi ;
fprintf('\nShoot at %f deg', thetaShoot) ;
```

15.5 SUMMARY

In this chapter, we discussed Real-Time Workshop and described two operational scenarios using Real-Time Workshop. First, we discussed using the Generic Real-Time Target. Next, we illustrated development of an xPC target. It was possible here to present only a brief introduction to Real-Time Workshop. Details on the many available targets can be found on The MathWorks Web site and from target vendors.

REFERENCES

[1] *Real-Time Workshop: Getting Started* (Natick, Mass.: The MathWorks, Inc., 2002). The Getting Started manual provides an overview of Real-Time Workshop and describes the build process in some detail. An online reference is also supplied with Real-Time Workshop, and provides more detail on general Real-Time Workshop topics.

[2] *xPC Target Getting Started Guide* (Natick, Mass.: The MathWorks, Inc., 2002). The Getting Started manual provides details on setting up xPC and overview of xPC programming. An online reference manual provides more detail.

Block Reference

This appendix presents brief descriptions of all blocks in the Simulink block library. Additional useful blocks may be found in the Simulink Extras block library and block libraries associated with toolboxes.

A.1 CONTINUOUS BLOCK LIBRARY

TABLE A.1: Continuous Block Library

Block Icon	Purpose
du/dt — Derivative	Compute the time rate of change of the input.
$\frac{1}{s}$ — Integrator	Compute the time integral of the input signal.
$x' = Ax+Bu$ $y = Cx+Du$ — State-Space	Model a linear time-invariant multiple-input, multiple-output system or subsystem using state-space notation.
$\frac{1}{s+1}$ — Transfer Fcn	Implement a continuous transfer function.
Transport Delay	Simulate a time delay. The output is the input delayed by a specified time.
Variable Transport Delay	Simulate a variable time delay. The first input is the signal to be delayed. The second input is the length of the delay. Thus, the length of the delay can change during a simulation.
$\frac{(s-1)}{s(s+1)}$ — Zero-Pole	Implement a continuous transfer function using zero-pole notation.

A.2 DISCONTINUITIES BLOCK LIBRARY

TABLE A.2: Discontinuities Block Library

Block Icon	Purpose
Backlash	Implement a backlash nonlinearity.
Coulomb and Viscous Friction	Implement a simple model of Coulomb and viscous friction.
Dead Zone	Model a dead zone, which is a region in which the output is zero. The upper and lower limits of the dead zone are configuration parameters. If the input is below the dead zone, the output is the input minus the lower limit, and if the input is above the dead zone, the output is the input minus the upper limit.
Hit Crossing	Compare input signal to a preset value. Output a value of zero except during the step the signal passes through the preset value or remains at the preset value after reaching it.
Quantizer	Model an analog to digital converter. Its output is a multiple of the quantitation interval, which is a block configuration parameter.
Rate Limiter	Limit the rate of change of the output signal. When the input is changing, the rate of change of the output will be the same as the rate of change of the input, as long as the rate of change of the input is less than a setable limit. If the rate of change of the input exceeds the limit, the rate of change of the output will be the same as the limit.
Relay	Simulate a relay. The output is one of two specified discrete values, depending on the value of the input.
Saturation	Implement a saturation nonlinearity. The upper and lower limits of the output signal are configuration parameters. If the value of the input signal is between the limits, the value of the output will be the same as the value of the input.

A.3   DISCRETE BLOCK LIBRARY

TABLE A.3: Discrete Block Library

Block Icon	Purpose
$\dfrac{1}{z+0.5}$ Discrete Transfer Fcn	Implement a discrete transfer function using the notation (polynomials of z) frequently associated with control systems.
$\dfrac{(z-1)}{z(z-0.5)}$ Discrete Zero-Pole	Model a discrete transfer function using zero-pole notation.
$\dfrac{1}{1+0.5z^{-1}}$ Discrete Filter	Implement a discrete transfer function using the notation (polynomials of z^{-1}) frequently associated with digital filtering.
y(n)=Cx(n)+Du(n) x(n+1)=Ax(n)+Bu(n) Discrete State-Space	Model a linear time-invariant multiple-input, multiple-output discrete system using state-space notation.
$\dfrac{T}{z-1}$ Discrete-Time Integrator	Compute a discrete approximation to a continuous integrator.
First-Order Hold	Provide a piecewise linear approximation of the block input. At an offset time δ since the last sample $[x(k)]$, the output is $x(k) + \dfrac{\delta}{T}[x(k) - x(k-1)]$, where T is the sample period.
Memory	Output the value of the block input at the beginning of the previous time step. Note that with variable-step size solvers, the duration of the delay will vary as the simulation progresses. If a constant memory delay is needed, use a Transport Delay block.
$\dfrac{1}{z}$ Unit Delay	Output the input signal delayed by one sample time.
Zero-Order Hold	Output the input at the most recent sample time.

A.4   LOOKUP TABLES BLOCK LIBRARY

TABLE A.4: Lookup Tables Block Library

Block Icon	Purpose
2–D T[k] Direct Lookup Table (n–D)	Perform table lookup.
2–D T[k,f] Interpolation (n–D) using PreLookup	Work with the Prelookup Index Search Block to perform n-dimensional interpolation.
Lookup Table	Linearly interpolate in a one-dimensional table with extrapolation at both ends.
LookUp Table (2–D)	Map two inputs to a single output.
2–D T(u) Lookup Table (n–D)	Map n inputs to a single output.
u k f PreLookup Index Search	Find the index of the value of the input signal in a range of numbers. This block can produce appropriate input for the Interpolation (n–D) Using Prelookup block.

A.5 MATH OPERATIONS BLOCK LIBRARY

TABLE A.5: Math Operations Block Library

Block Icon	Purpose
Abs	Compute the absolute value of each component of the input signal, which may be scalar or vector.
Algebraic Constraint	Enable a Simulink model to solve algebraic equations. The model must be configured such that the input to the Algebraic Constraint block is dependent upon the value of the output. A model containing an Algebraic Constraint block will attempt to adjust the value of the block output such that the value of the block input is 0.
Assignment	Assign values to selected elements of the input signal.
Bitwise Logical Operator	Perform bitwise logical operations (AND, OR, XOR, etc.) on integer input signal with a specified operand.
Combinatorial Logic	Look up the elements of the input vector in a truth table.
Complex to Magnitude-Angle	Accept a complex input signal. Output can be configured as the phase angle, magnitude, or both.
Complex to Real-Image	Accept a complex input signal. Output can be configured as the real part, the imaginary part, or both.
Dot Product	Accept two vector signals of the same dimension. The output is the dot product of the current input vectors.
Gain	Multiply the block input by a constant. The Gain block will work with scalar or vector signals, and the value of gain may be a scalar or vector compatible with the input signal.
Logical Operator	Implement a number of logical operations such as AND and OR.

TABLE A.5: Math Operations Block Library (Cont)

Block Icon	Purpose
Magnitude-Angle to Complex	Accept magnitude, phase angle, or both as input and output corresponding complex signal.
Math Function	Perform a variety of mathematical functions such as exp, log, and square root.
Matrix Concatenation	Concatenate matrix signals.
Matrix Gain	Multiply the block input vector by a compatible matrix. The input vector is treated as a column vector. Therefore, the gain matrix must have the same number of columns as there are elements in the input vector. The number of outputs is the same as the number of rows in the gain matrix, which does not have to be square.
MinMax	Compute the minimum or maximum value of the current block inputs. The number of input ports can be set in the block dialog box.
Polynomial	Implement MATLAB function `polyval`.
Product	The Product block can be configured with one or more inputs. If there is one input, the output is the product of all elements of the input vector. If there are multiple inputs, the output is the element-by-element product of the vectors at each input port.
Real-Image to Complex	Accept real and imaginary components and output complex signal.
Relational Operator	Implement a number of relational operations such as less than or equal, greater than, etc.
Reshape	Change the dimensions of the input matrix signal. Implement MATLAB function `reshape`.

TABLE A.5: Math Operations Block Library (Cont)

Block Icon	Purpose
floor Rounding Function	Perform a variety of rounding operations including round, round up (ceil), round down (floor), and round to the nearest integer toward zero (fix).
Sign	Implement the signum nonlinearity. The output is 1 if the input is positive, 0 if the input is 0, and −1 if the input is negative.
1 Slider Gain	Allow the user to set a gain using a slider control. Open the slider control by double-clicking the block. The slider can be moved during a simulation, thus providing a variable input device.
(sum)	Compute the algebraic sum of the block inputs. The number of inputs and the sign applied to each input can be set in the block dialog box.
sin Trigonometric Funtion	Implement standard trigonometric functions such as sine and cosine.

A.6   PORTS & SUBSYSTEMS BLOCK LIBRARY

TABLE A.6: Ports & Subsystems Block Library

Block Icon	Purpose
Template Configurable Subsystem	Place in a block library to produce a configurable subsystem block.
In1 Out1 Atomic Subsystem	Template for atomic subsystem.
Enable	Convert a subsystem into an enabled subsystem.
In1 Out1 Enabled Subsystem	Template-enabled subsystem.

TABLE A.6: Ports & Subsystems Block Library (Cont)

Block Icon	Purpose
Enabled and Triggered Subsystem (In1, Out1)	Template-enabled and triggered subsystem.
For Iterator Subsystem (In1, for { ... }, Out1)	Template subsystem with For Iterator block.
Function-Call Generator (f())	Execute a function call subsystem.
Function-Call Subsystem (Function(), In1, Out1)	Template subsystem with trigger block configured to trigger on function call.
If (u1, if(u1 > 0), else)	An If block controls one or more If Action subsystems.
If Action Subsystem (Action, In1, Out1)	Conditional subsystem executed if commanded by If block.
In1 (1)	Create an input for a subsystem. An Inport block can also be used to receive an external input (for example, using the sim command) to a model.
Out1 (1)	Create an output port for a subsystem. An Outport block can also be used to produce model outputs, to be used, for example, by the linearization or trim commands.
Subsystem (In1, Out1)	Template subsystem.
Switch Case (u1, case [1]:, default:)	A Switch Case block is the decision-making block that controls Switch Case Action subsystems.

TABLE A.6: Ports & Subsystems Block Library (Cont)

Block Icon	Purpose
Switch Case Action Subsystem	Used in conjunction with Switch Case block. Each Switch Case output controls one Switch Case subsystem.
Trigger	Convert a subsystem into a triggered subsystem. Convert an enabled subsystem into a trigger when enabled subsystem.
Triggered Subsystem	Template triggered subsystem.
While Iterator Subsystem	Template subsystem containing a While Iterator block.

A.7 SIGNAL ATTRIBUTES BLOCK LIBRARY

TABLE A.7: Signal Attributes Block Library

Block Icon	Purpose
Data-Type Conversion	Convert input signal to a specified data type.
IC	Set the initial condition of its output to a specified value. After the simulation begins, the block output is the same as its input. This block is useful in algebraic loops, as it can provide an initial guess to the algebraic loop solver for the first time step.
Probe	Display a variety of useful information about the block input.
Rate Transition	Synchronizes transfer of data from block of one data rate to a block of different data rate.
Signal Specification	Implements error checking on input signal. If no errors are found, the signal is transmitted to the output unchanged. If an error is found, the simulation is forced to stop.
Width	Output the number of elements in the input vector. Thus, if the input is a five-element vector, the output is 5.

A.8 SIGNAL ROUTING BLOCK LIBRARY

TABLE A.8: Signal Routing Block Library

Block Icon	Purpose
	Groups multiple signal lines into a single data bus.
	Create a new signal containing a specified subset of the components of a data bus produced using a Mux block or Bus Creator block.
Data Store Memory	A Data Store Memory is a named memory location written to by Data Store Write blocks and read from by Data Store Read blocks. A simulation can save data in a Data Store Memory then access that data later.
Data Store Read	Output the current value of the contents of the corresponding Data Store.
Data Store Write	Write to a specified Data Store Memory block. More than one Data Store Write block can write to a particular Data Store Memory, but if two or more Data Store Write blocks attempt to write to the same Data Store Memory on the same simulation step, the results are unpredictable.
	Split a vector input signal into a configurable number of scalar output signals.
From	A From block works with a Goto block. The output of a From block is the same as the input to the corresponding Goto block. A From block can receive input from only one Goto block, but a Goto block can send a signal to any number of From blocks.
Goto	Send the block input to all corresponding From blocks.
Goto Tag Visibility	The Goto Tag Visibility determines which subsystems can contain From blocks corresponding to a particular Goto block.
Manual Switch	Switch between the two inputs when the block is double-clicked.

TABLE A.8: Signal Routing Block Library (Cont)

Block Icon	Purpose
Merge	Output is the most recently updated value of inputs. Most commonly driven by subsystems that execute alternately.
Multiport Switch	Accept a specified number of inputs. A control signal determines which input is passed to the output.
	Combines a configurable number of scalar input signals to produce a vector output signal.
Selector	Accept a vector input, and produce a vector output that consists of selected elements of the input vector in a selected order.
Switch	Switch between two input signals based on the value of a control signal.

A.9 SINKS BLOCK LIBRARY

TABLE A.9: Sinks Block Library

Block Icon	Purpose
Display	Display the current value of the input signal.
Floating Scope	Display scalar or vector signals in a method analogous to an oscilloscope. A Floating Scope receives its input signal from any signal line selected during model execution.
Out1	Create an output port for a subsystem. An Outport block can also be used to produce model outputs, to be used, for example, by the linearization or trim commands.
Scope	Display scalar or vector signals in a method analogous to an oscilloscope.
STOP Stop Simulation	Cause the simulation to stop when the input signal is nonzero.

TABLE A.9: Sinks Block Library (Cont)

Block Icon	Purpose
Terminator	Connect unused block outputs to Terminator blocks to prevent Simulink from producing error messages. For example, if you use a Demux block to split a vector signal but only need to use one component of the vector signal, connect the unneeded output ports of the Demux block to Terminator blocks.
untitled.mat / To File	Save the input signal to a file in MATLAB.mat format. The signal may be scalar or vector.
simout / To Workspace	Store the input signal in a MATLAB matrix accessible in the MATLAB workspace after the simulation stops. The signal may be scalar or vector.
XY Graph	Produce a graph using two scalar inputs. The signal connected to the top input port is the independent variable (x-axis), and the signal connected to the lower input port is the dependent variable (y-axis).

A.10   SOURCES BLOCK LIBRARY

TABLE A.10: Sources Block Library

Block Icon	Purpose
Band-Limited White Noise	Generate a signal containing band-limited white noise of a specified power spectral density (PSD).
Chirp Signal	Generate a sinusoidal signal of continuously increasing frequency.
Clock	Generate a signal consisting of the current simulation time.
1 / Constant	Generate a constant value. The constant can be a scalar or a vector.
12:34 / Digital Clock	Generate a signal consisting of the current simulation time, sampled at a specified period. Equivalent to a combination of a Clock block and a Zero-Order hold but much more efficient.

TABLE A.10: Sources Block Library (Cont)

Block Icon	Purpose
simin From Workspace	Generate a signal by interpolating in a table defined by variables in the MATLAB workspace.
untitled.mat From File	Generate a signal by interpolating in a MATLAB matrix stored in a file.
Ground	Connect Ground blocks to unused inputs to prevent Simulink from producing error messages. For example, if a State Space block is used to model the unforced behavior of a system, connect a Ground block to its input port.
1 In1	Create an input for a subsystem. An Inport block can also be used to receive an external input (for example, using the sim command) to a model.
Pulse Generator	Generate a rectangular wave. Configuration parameters are period, amplitude, duty cycle, and start time.
Ramp	Generate a signal for which the time derivative is a constant.
Random Number	Generate a signal containing normally distributed random numbers.
Repeating Sequence	Generate an arbitrary periodic signal. The signal is defined by a table of time points and amplitudes.
Signal Generator	Generate a periodic signal (sine wave, square wave, or sawtooth wave) or random noise. Configuration parameters are signal amplitude and frequency.
Signal 1 Signal Builder	Produce a signal defined by using a graphical user interface. Double-click the block to open a graphical signal editor.
Sine Wave	Generate a sine wave. Amplitude, phase, and frequency can be set.
Step	Generate a step function. Configuration parameters are step time, initial value, and final value.
Uniform Random Number	Generate a signal containing uniformly distributed random numbers.

A.11   USER-DEFINED FUNCTIONS BLOCK LIBRARY

TABLE A.11: User-Defined Functions Block Library

Block Icon	Purpose
f(u) Fcn	Implement a function using a C language syntax. This block can accept a vector input but produces a scalar output. This block cannot perform matrix arithmetic; however, it is faster than the MATLAB Fcn block, and is therefore preferable when matrix arithmetic is not needed.
MATLAB Function MATLAB Fcn	Implement a function using MATLAB syntax. It can accept vector inputs and produce vector outputs.
system S-Function	Incorporate a block written in MATLAB or C code into a Simulink model.
system S-Function Builder	Open a graphical user interface that helps automate the coding of C S-Functions.

B

Parameter Reference

Listed in this appendix are parameters that may be set using the set_param command and read using the get_param command. The command syntax to read a parameter is

```
get_param(path, parameter_name) ;
```

where *path* is a MATLAB string containing the path to the object, and *parameter_name* is a MATLAB string containing the parameter name from this appendix. For a model, *path* is the model name. For a block, *path* is the path to the block, starting with the model name. For example, the path to a Gain block in system example would be 'example/gain'. The syntax of the set_param command is

```
set_param(path, parameter_name, parameter_value) ;
```

Here, *parameter_value* is a string or numeric expression, depending on the particular parameter. Almost all parameter values are entered as strings. For example, to set the value of gain for the Gain block in system example to 5.0, the following statement could be used:

```
set_param('example/gain','Gain','5.0') ;
```

An important exception to this rule is the UserData parameter for models, masked subsystems, and blocks. UserData can accept any MATLAB variable, including arrays, cell arrays, and structures.

B.1 MODEL PARAMETERS

Table B.1 lists parameters that are defined for a model as a whole.

TABLE B.1: Model Parameters

Parameter	Purpose
AbsTol	Solver absolute tolerance
AlgebraicLoopMsg	Setting for algebraic loop diagnostic: none, warning, error
ArrayBoundsChecking	Setting for array bounds checking diagnostic: none, warning, error
AutoZoom	Printing auto-zoom option: on, off

TABLE B.2: Model Parameters (Cont)

Parameter	Purpose
BlockAttributesDataTip	Setting of block attributes data tips: on, off
BlockDataTips	Setting of block data tips: on, off
BlockDescriptionStringDataTip	Setting to activate the user description string data tip for all blocks that have nonempty user description strings: on, off
BlockMaskParametersDataTip	Setting to activate the block mask parameters data tips: on, off
BlockParametersDataTip	Setting to activate the block parameters data tip: on, off
BlockPortWidthsDataTip	Setting to activate the block port width data tip: on, off
BlockReductionOpt	Setting to control block reduction optimization: on, off
BooleanDataType	Enable Boolean type checking: on, off
BrowserLookUnderMasks	Configure model browser to look under masks: on, off
BrowserShowLibraryLinks	Configure model browser to show library links: on, off
BufferReuse	Enable buffer reuse for storage optimization: on, off
CheckForMatrixSingularity	Matrix singularity checking: none, warning, error
ConfigurationManager	Name of configuration manager software: None, RCS, Microsoft Visual SourceSafe, PVCS
ConsistencyChecking	Consistency checking: none, warning, error
CovHtmlReporting	Generate HTML coverage report: on, off
CovNameIncrementing	Enable incrementing the coverage report variable name each time a model is run: on, off
CovPath	Model path for coverage instrumentation
CovSaveName	Name of MATLAB variable in which to save coverage information
Created	MATLAB string containing the time the model was created, in the format *ddd mmm nn hh:mm:ss yyyy*, e.g., Tue Jul 04 23:04:45 1776
Decimation	Decimation factor
ExternalInput	Variable names for time and input variables if LoadExternalInput is on
FinalStateName	Variable name in which to save final state in the MATLAB workspace

TABLE B.1: Model Parameters (Cont)

Parameter	Purpose
FixedStep	Stepsize for fixed-step size solvers
InheritedTsInSrcMsg	Setting of diagnostic for sample time of -1 in source blocks: none, warning, error
InitialState	Variable name from which to load initial state
InitialStep	Initial integration step size
Int32ToFloatConvMsg	Setting of diagnostic message for 32-bit integer to floating-point-type conversion: none, warning, error
IntegerOverflowMsg	Setting of diagnostic message for integer overflow: none, warning, error
InvariantConstants	Simulation parameter to disable sampling of blocks that produce a constant output, such as the Constant block: on, off. Setting this flag to on makes it impossible to change a constant value during a simulation and also speeds up the simulation somewhat.
LastModifiedDate	MATLAB string containing the time the model was last saved modified, in the format *ddd mmm nn hh:mm:ss yyyy*, e.g., Tue Jul 04 23:04:45 1776
LimitMaxRows	Limit the number of rows in model output variables: on, off
LinearizationMsg	Setting of diagnostic message for linearization: none, warning, error
LoadExternalInput	Load input from MATLAB workspace: on, off
LoadInitialState	Load initial state from MATLAB workspace: on, off
Location	Model window location and size vector
MaxOrder	Maximum order for solver ode15s
MaxRows	Maximum number of rows in output
MaxStep	Maximum integration step size
MinStepSizeMsg	Setting for minimum step size violation messages: none, warning, error
ModelBrowserVisibility	Setting of flag to display the model browser pane in the model window: on, off
ModelBrowserWidth	Width of the model browser pane in pixels
ModelVersionFormat	Format for **Model version format** field in the **File:Model properties** dialog box **Options** page. This should be set for compatibility with a configuration control program such as PVCS. The default value is 1.%<AutoIncrement:2>

TABLE B.1: Model Parameters (Cont)

Parameter	Purpose
ModifiedByFormat	Format for **Modified by** format field in the **File:Model** properties dialog box **Options** page. This should be set for compatibility with a configuration control program such as PVCS. The default value is %<Auto>
ModifiedDateFormat	Format for **Modified date** format field in the **File:Model properties** dialog box **Options** page. This should be set for compatibility with a configuration control program such as PVCS. The default value is %<Auto>
MultiTaskRateTransMsg	Setting of the **MultiTask Rate Transition** configuration option (**Simulation: parameters** dialog box): none, warning, error
Name	Model name. This is normally the filename without the extension.
Open	Indicates whether system is open or closed: on, off. Setting Open to off closes the system, setting Open for a closed system to on opens the system
OptimizeBlockIOStorage	Setting of the **Simulation:parameters** block dialog option Optimize block I/O storage: on, off
OutputOption	Output option from list. AdditionalOutputTimes, RefineOutputTimes, Specified
OutputSaveName	Variable name in which to save output in the MATLAB workspace
OutputTimes	Vector of output times depending on setting of OutputOption
PaperOrientation	Paper orientation for printing the model: portrait, landscape
PaperPositionMode	Paper position mode: auto, manual. Set to manual to explicitly specify the figure position on the page.
PaperType	Printer configuration parameter **Paper type**, options depend on printer selected
PaperUnits	Printer configuration units parameter: inches, mm, points
Profile	Generate a model profile file when the model is executed: on, off. The profiler output is stored in model_profile.html, where model is the model name. The profiler output can be viewed using a Web browser.

TABLE B.1: Model Parameters (Cont)

Parameter	Purpose
Refine	Refine factor
RelTol	Relative tolerance
SampleTimeColors	Sample time colors: on, off
SaveFinalState	Save final simulation state vector: on, off
SaveFormat	Format of simulation output to the MATLAB workspace: Matrix, Structure, StructureWithTime
SaveOutput	Save simulation output: on, off
SaveState	Save state trajectory: on, off
SaveTime	Save simulation time: on, off
SfunCompatibilityCheckMsg	Setting of S-function upgrades needed diagnostic: none, warning, error
ShowLineWidths	Display the dimension of each vector signal line: on, off
ShowPortDataTypes	Display the data type of each output port: on, off
SignalLabelMismatchMsg	Setting of signal level mismatch diagnostic: none, warning, error
SimParamPage	Name of the page on which to open the **Simulation:Parameters** dialog box. Choices are Solver, Workspace I/O, Diagnostics, Advanced, and if Real-Time Workshop is installed, RTW
SingleTaskRateTransMsg	Setting of SingleTask rate transition diagnostic: none, warning, error
Solver	Solver name: ode45, ode23, ode113, ode15s, ode23s, ode5, ode4, ode3, ode2, ode1, discrete
SolverMode	Mode setting for fixed-step size solvers: SingleTasking, MultiTasking, Auto
StartTime	Simulation start time
StateSaveName	Variable name in which to save the model state trajectory in the MATLAB workspace
StatusBar	Status bar visibility: on, off
StopTime	Simulation stop time
TimeSaveName	Variable name in which to save time in the MATLAB workspace
ToolBar	Toolbar visibility: on, off
UnconnectedInputMsg	Setting for unconnected input messages: none, warning, error
UnconnectedLineMsg	Setting for unconnected signal line messages: none, warning, error

TABLE B.1: Model Parameters (Cont)

Parameter	Purpose
UnconnectedOutputMsg	Setting for unconnected output messages: none, warning, error
UnnecessaryDatatypeConvMsg	Setting of unneeded type conversions diagnostic: none, warning, error
UpdateHistory	Setting of model update history option: UpdateHistoryNever, UpdateHistoryWhenSave
VectorMatrixConversionMsg	Setting of Vector/Matrix conversion diagnostic: none, warning, error
Version	Simulink version used to modify the model the last time it was saved
WideVectorLines	Wide vector lines set to on, off
ZeroCross	Zero crossing detection: on, off
ZoomFactor	Zoom factor in percent

B.2 COMMON BLOCK PARAMETERS

Table B.2 lists parameters that are common to all blocks.

TABLE B.2: Common Block Parameters

Parameter	Purpose
Name	Block name
Parent	Name of object that owns the block
BlockType	Block type. This is the name in the top left corner of the block dialog box (not the block dialog box title bar).
BlockDescription	Text in description field of block dialog box
InputPorts	Array listing the locations of the block input ports
OutputPorts	Array listing the locations of the block output ports
Orientation	Block orientation: right, left, down, up
ForegroundColor	Block foreground color: black, white, etc.
BackgroundColor	Block background color: black, white, etc.
DropShadow	Drop shadow: on, off
NamePlacement	Placement of block name: normal, alternate
FontName	Name of font used on block
FontSize	Size of font in points
FontWeight	Relative weight of font: light, demi, bold
FontAngle	From list normal, italic, oblique
Position	Location of block in model window: left, top, right, bottom. This parameter must be entered as a vector, rather than as a string.
ShowName	Show the block name: on, off

TABLE B.2: Common Block Parameters (Cont)

Parameter	Purpose
Tag	Block tag. This is a user-defined variable. It must be a character string if used.
UserData	Any MATLAB variable. Note that UserData is not saved with the model.
Selected	Selection state of block: on, off. Note that Selected is not saved with the model.
SourceBlock	If BlockType is Reference, this is the corresponding library block.
SourceType	If BlockType is Reference, this is the corresponding library block BlockType.

B.3   MASK PARAMETERS

Table B.3 lists parameters that are specific to masked blocks.

TABLE B.3: Masked Block Parameters

Parameter	Purpose
MaskType	Contents of field **Mask Type**
MaskDescription	Contents of **Block Description**
MaskHelp	Contents of **Block Help**
MaskPromptString	Contents of **Prompt** field. If there are multiple prompts, they are separated with the | symbol.
MaskStyleString	Type field: Edit, Checkbox, Popup. If there are multiple prompts, they are separated with |.
MaskVariables	**Variable** field. If there are multiple prompts, the corresponding variable names are separated with |.
MaskInitialization	Contents of **Initialization commands**
MaskDisplay	Contents of **Drawing commands**
MaskIconFrame	Icon frame: on, off
MaskIconOpaque	Icon transparency: on, off
MaskIconRotate	Icon rotation: on, off
MaskIconUnits	**Drawing coordinates**: Pixel, Autoscale, Normalized
MaskValueString	Contents of the masked block's dialog box fields, in the form of a string with the field contents separated with |.

B.4   IDENTIFYING BLOCK PARAMETERS

In addition to the common block parameters discussed in Section B.2, most blocks have additional parameters corresponding to block dialog box fields. To identify the parameter name associated with a block, create a model containing only the block. Save the model and then open the .mdl file using the MATLAB editor. Scroll down to locate the block parameters. To illustrate the process, suppose we wish to set the gain value of a Gain block.

Create a new model with a single Gain block. Save the model. Here we've saved the model as `GainParams.mdl`.

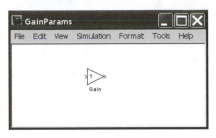

Open the model with the MATLAB editor (or any text editor). Scroll down to the block structure. The Gain block section of our file `GainParams.mdl` is shown in Listing B.1.

Listing B.1: `GainParams.mdl`

```
BlockParameterDefaults {
  Block {
    BlockType               Gain
    Gain                    "1"
    Multiplication          "Element-wise(K.*u)"
    ShowAdditionalParam     off
    ParameterDataTypeMode   "Same as input"
    ParameterDataType       "sfix(16)"
    ParameterScalingMode    "Best Precision: Matrix-wise"
    ParameterScaling        "2^0"
    OutDataTypeMode         "Same as input"
    OutDataType             "sfix(16)"
    OutScaling              "2^0"
    LockScale               off
    RndMeth                 "Floor"
    SaturateOnIntegerOverflow on
  }
}
```

Index